FROM THE FILMS OF

Harry Potter

HERBOLOGY MAGIC

FROM THE FILMS OF

Harry Potter

HERBOLOGY MAGIC

*Botanical Projects, Terrariums, and Gardens
Inspired by the Wizarding World*

CRAFTS BY
JIM CHARLIER

TEXT BY
JODY REVENSON

**Titan
Books**

CONTENTS

6   *Introduction*

Chapter 1: Terrariums

12   Terrarium Terminology

15   Herbology Classroom Wardian Case

21   Sybill Trelawney's Divination Classroom Terrarium

27   Potions Classroom Terrarium

33   Mad-Eye Moody's Flask Terrarium

39   Argus Filch's Lantern Terrarium

45   Carnivorous Plant Bog Garden

51   Black Lake Water Garden

57   Luna Lovegood's Living Necklace

Chapter 2: Indoor Décor

63   Sprout's Intro to Herbology Garden

67   Kitchen Herbology

71   Portkey Planters

75   Cornish Pixie Air Plant Cage

81   Hogwarts Houses Macramé Plant Hangers

87   Neville Longbottom's Propagation Station

93   Whomping Willow Bonsai Tree

Chapter 3: Outdoor Décor

- 101 Professor Flitwick's Charmed Ever-Raining Rain Chain
- 107 Wizarding Wayfinding Signpost
- 113 Deathly Hallows Trellis
- 119 Garden of Prophecy Trellis Orbs
- 123 Sorting Hat Hanging Basket
- 129 Forbidden Forest Fence Planter
- 135 Dirigible Plums Sign
- 141 Privet Drive House Numbers
- 145 Pensieve Bird and Bee Bath
- 149 Great Hall Floating Candle Lights
- 153 Trevor the Toad's Abode

Chapter 4: Every Garden Is Magic

- 159 Wizarding World Plant Substitutes
- 162 Real but Weird Plants
- 166 Types of Gardens
- 168 The Magic of Pollinators
- 169 Why Use Latin for Spells, Charms, and Plant Names?
- 170 Zone Grown
- 171 Native Plants

INTRODUCTION

"You know, if you're interested in plants, you'd be better off with 'Goshawk's Guide to Herbology.' Do you know there's a wizard in Nepal who's growing gravity-resistant trees?"

—NEVILLE LONGBOTTOM
HARRY POTTER AND THE GOBLET OF FIRE

Herbology, a core subject at Hogwarts School of Witchcraft and Wizardry, is the study of magical and non-magical plants used for potions such as Polyjuice Potion (with knotgrass and fluxweed), Felix Felicis (utilizing tincture of thyme and powdered common rue for luck), and medicines such as Skele-Gro (which contains fanged geranium). Plants are also employed for individual and unique uses; for example, Neville Longbottom gave Harry Potter Gillyweed to complete the second task under the waters of the Black Lake in *Harry Potter and the Goblet of Fire*. Gillyweed allows the consumer to grow gills, fins, and flippers. Mandrakes are used in the Mandrake Restorative Draught, which cures those who have been Petrified. Cultivating and using the Mandrakes grown in the greenhouses at Hogwarts to create the draught is crucial to the events in *Harry Potter and the Chamber of Secrets*. Luckily, Professor Pomona Sprout's first lesson to her second-year students was all about repotting these constantly growing plants.

Harry Potter: Herbology Magic contains ideas for making your garden as magical as a stroll through Greenhouse Three or a foray into the Forbidden Forest. The Wizarding World offers a diverse range of foliage, from plants that charm to plants that shock, and your garden can employ facsimiles that will evoke Dirigible Plums and Venomous Tentacula within enchanting displays.

Locations for your garden or greenhouse may also influence the types of plants you choose and how they are arranged. Even if you live in a place as small as the Cupboard Under The Stairs, there are plants that can enhance it with minimal care. You can showcase a Cornish Pixie surrounded by air plants or have Mad-Eye Moody's eye watching from a terrarium shaped like a flask. Terrariums are a great boon to the budding herbologist, being fully enclosed glass containers to cultivate moisture-loving plants indoors.

You can "plant" a Wizarding Wayfinding Signpost in an outdoor garden, pointing to special wizarding locations, or listen to the pitter-patter of rain sounds with Professor Flitwick's Charmed Ever-Raining Rain Chain. There are trellises in the shapes of the Deathly Hallows and Hall of Prophecy Orbs that will support climbing plants and vines. And no

wizarding-style garden would be complete without Portkey Planters or a Pensieve Bird and Bee Bath.

There are templates for a few of these projects available at **www.InsightEditions.com/HarryPotterHerbologyMagic**. And we definitely encourage sharing your projects with other witches, wizards, and even Muggles at **#harrypotterherbology**.

Harry Potter: Herbology Magic contains almost as many crafts and activities as books in Hermione Granger's beaded bag (with its Undetectable Extension Charm) with step-by-step instructions that will enhance your home, lawn, or landscape with bewitching planters and beguiling trellises. So, don't beat around the Dirigible Plums bush—we've laid the groundwork down here for you to dig in and make magic with Herbology!

Safety Tips

Hand and power tools can be more dangerous than a Bludger on a Quidditch pitch. In order to avoid accidental splinching, follow the below guidelines:

- Most projects herewith are not suitable for children alone. Adult supervision is required.

- Only use tools you know how to use properly and safely.

- For working with wood and wire, safety glasses are necessary. Projects marked with a ⌒ symbol require safety glasses.

- When working with wire you will find leather gloves invaluable. Projects marked with a ✋ symbol require leather safety gloves.

- Any time you work with sandpaper or a sander, a mask is required. Projects marked with a 😷 symbol require a mask.

- Wear ear protection when working with power tools. Some of those tools are as loud as an unpotted Mandrake. Projects marked with a 🎧 symbol require ear protection.

Chapter 1

TERRARIUMS

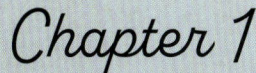

"You grasp your Mandrake firmly. You pull it sharply up out of the pot. Got it. And now you dunk it down into the other pot and pour a little sprinkling of soil to keep him warm."

—PROFESSOR POMONA SPROUT,
HARRY POTTER AND THE CHAMBER OF SECRETS

TERRARIUM TERMINOLOGY

1. Terrarium Layering

Layer depths will depend on terrarium height and width.

2. Décor Layer

Any sand, coarse gravel, art, stones, driftwood, sticks, shells, glass, preserved moss, etc.

3. Growing Medium Layer

THICKEST LAYER–POTTING MIX

This is the layer in which the plants grow. Plants with similar needs will always be more successful. Different types of plants will need different soil mixes:

- Leafy Terrarium Plants: houseplant soil mix
- Cactus & Succulents: porous, coarse-sand potting mix
- Carnivorous Plants: porous, low-nutrient, acidic soil mix

4. Charcoal Layer

THIN LAYER–ACTIVATED HORTICULTURAL CHARCOAL

Charcoal keeps moisture in a terrarium; absorbs excess water; helps eliminate odors; and guards against algae, fungus, rot, and bacteria. This level is sometimes placed on top of the drainage layer.

5. Barrier Layer

THINNEST LAYER–SPHAGNUM MOSS OR A NYLON OR METAL MESH SCREEN

This layer keeps the potting soil and water in the drainage layer from mixing and making mud, while enabling water to pass through to the drainage layer.

6. Drainage Layer

1-INCH LAYER FOR MOST TERRARIUMS–LAVA ROCK

This is a reservoir area at the bottom of a terrarium for excess water. It enables the plants to grow above without having their roots sitting in water. Sand is often mixed with lava rocks.

Herbology Classroom Wardian Case

*"Today, we're going to repot Mandrakes.
Who here can tell me the properties of the Mandrake root?"*

—PROFESSOR POMONA SPROUT,
HARRY POTTER AND THE CHAMBER OF SECRETS

SKILL LEVEL: 3
TIME: 2 TO 3 HOURS

Terrariums are one of the easiest, most versatile, and creative ways to cultivate plants in your own home. A Wardian case is one of the earliest forms of terrariums, created by an English botanist in the mid-1800s in an effort to import newly discovered flora species from Europe over long sea voyages. The 1800s was also when Temperate House at Kew Royal Botanic Gardens was built, which was the inspiration for Greenhouse Three set design, where Professor Sprout teaches Herbology in *Harry Potter and the Chamber of Secrets*. Temperate House is the largest existing Victorian glass greenhouse. Greenhouse Three did not need to be as large but did need to be long, as production designer Stuart Craig wanted to emphasize the table where the students are taught how to repot Mandrakes. Greenhouse Three's walls were lined with large-leafed plants in pots, and its table was covered with fifty radio-controlled animatronic baby Mandrakes that squirmed and screamed, to the annoyance of anyone not wearing earmuffs. No earmuffs needed here!

What you need

- Terrarium windowed box, with doors (project uses 10-by-15-inch terrarium box)
- Acrylic tray, a length and width to fit comfortably inside terrarium box, it should be 2 to 3 inches in height (project uses 9-by-12-inch acrylic tray)
- Lava rock
- Sand
- Scissors
- Sphagnum moss or optional nylon or metal mesh screen
- Horticultural charcoal
- Potting soil
- Tile(s) or flat rock
- Plants (see Plants below)
- Tiny terra-cotta pot
- Tiny terrarium jars
- Preserved moss or screen
- Dollhouse furniture and garden items, such as a large table, etc.

Instructions

1. Add a layer of lava rock and sand to the acrylic tray.
2. Add a ½-inch layer of sphagnum moss, or cut a screen to fit on top of the lava rocks. This will keep soil from mixing with the lava rocks which will form a layer for water to sit if the terrarium is overwatered, keeping roots out of constant wetness.
3. Place the horticultural charcoal on top of the screen or moss. It doesn't have to be an entire layer. The charcoal will help with moisture retention and mildew control. Just a scattering should be fine.
4. Add a layer of the potting soil, up to just about the rim of the acrylic tray.
5. Place the tile(s) or flat stone in the middle of the tray to form the floor of the classroom.
6. Add plants around the tile floor. Keep taller plants toward the "back" of the setting. The back will depend on which direction most will view the terrarium.
7. Plant up a tiny terra-cotta pot and a tiny terrarium jar or two with terrarium plants or air plants.
8. Add some puffs of preserved moss to add some instant color. The preserved moss will also help retain moisture for the rest of the live plants.
9. Add dollhouse furniture—a large table and other garden-oriented implements.
10. Place the acrylic tray inside the terrarium.

Plants

Any houseplant store will have a good selection of terrarium-suitable plants. Ask a plant store professional for recommendations. You can find collections of different terrarium plants available online. Some suggested plants:

- *Ficus pumila* 'Quercifolia' (string of frogs, miniature oakleaf fig)
- *Fittonia albivenis* (nerve plant)
- *Nephrolepis exaltata* 'Fluffy Ruffles' (sword fern, dwarf Boston fern)
- *Senecio rowleyanus* (string of pearls)
- *Hatiora salicornioides* (dancing-bones cactus)
- *Thuidium delicatulum* (common fern moss)
- *Peperomia caperata* 'Rosso' (emerald ripple peperomia, radiator plant)
- *Peperomia prostrata* (string of turtles)
- *Pilea microphylla* (artillery fern)
- *Pilea cadierei* (aluminum plant)
- *Hypoestes phyllostachya* (polka dot plant)
- *Soleirolia soleirolii* (baby's tears)
- *Pellaea rotundifolia* (button fern)
- *Tillandsia* (air plants)

Care

This project is an open terrarium. Technically, it is plants sharing a large, shallow pot.

- *Light:* This terrarium will need bright yet indirect light. North-facing windows are generally best for this type of light. They will perform best with 8 to 10 hours of light per day. A full spectrum grow light may be necessary. Plants grow toward light, so light from above is ideal.

- *Water:* Since there are no drain holes in the acrylic tray, it will be easy to overwater. Only water lightly once a week or when the soil feels dry to the touch, or plants show signs of droopy leaves. Watering with a mister (spray bottle) is best. Terrarium plants do like humidity, so a mist between waterings will make them happy. If leaves seem to be turning brown and getting crispy edges, they need more humidity.

- *Maintenance:* As plants grow, they need to be trimmed and even replaced if they get too large, or crowd out other plants. Always trim dead leaves to keep the terrarium looking its best.

SYBILL TRELAWNEY'S DIVINATION CLASSROOM TERRARIUM

"Broaden your mind! You must look beyond. The art of crystal-gazing is in the clearing of the inner eye. Only then can you see!"

—SYBILL TRELAWNEY,
HARRY POTTER AND THE PRISONER OF AZKABAN

SKILL LEVEL: 3
TIME: 2 TO 3 HOURS

In addition to tessomancy, the art of reading tea leaves, Divination Professor Sybill Trelawney teaches crystal-gazing, a method for viewing the future based on visions seen in a crystal ball. Harry and Ron struggle with their crystal ball (actually, Ron is snoring). Hermione has little interest in Divination, and she purposefully knocks the ball off the table, where it disappears down the stairs. After Harry returns it to the classroom, Professor Trelawney makes only the second accurate prediction in her career—concerning the return of Peter Pettigrew, who betrayed Harry's parents to Voldemort, leading to their deaths.

Not every vision into a crystal ball is negative or dark. Looking inside this crystal ball-shaped terrarium with its tiny version of a red-covered table and chairs from the Divination class set design, surrounded by exotic succulents and puffs of mini mosses, is a most enchanting vision.

What you need

- 12-inch-round, crystal-ball-shaped, open terrarium (could also be a fish bowl)
- Lava rock
- Sand
- Scissors
- Sphagnum moss or optional nylon or metal screen
- Horticultural charcoal
- Cactus and succulent potting soil
- Terra-cotta planter saucer
- Terrarium plant tongs, shovel, and brush (or paintbrush)
- Plants (see plants on page 24)
- Glue gun
- Small terra-cotta pot
- Red fabric
- Small cork
- Clear marble
- Colored beach glass, crystals, or amethyst
- Champagne Cage Chairs (see page 25)
- Preserved moss

Instructions

1. Add a layer of lava rock and sand to the bottom of the terrarium—an inch should do.
2. Add a ½-inch layer of sphagnum moss, or a screen, on top of the lava rocks. This will keep soil from mixing with the lava rocks which will form a layer for water to sit on if the terrarium is overwatered, keeping roots out of constant wetness.
3. Place a layer of horticultural charcoal on top of the screen or moss. The charcoal will help with moisture retention and mildew control.
4. Add 2 to 3 inches of the potting soil.
5. Add the terra-cotta plant saucer in the center of the terrarium, upside down, as the base for table and chairs.
6. Plant up the terrarium with 7 to 8 succulent plants keeping the tallest in the back. Use the terrarium plant tongs and shovel to position each plant in place.
7. Add a light, decorative, layer of sand over the soil, around all the plants. You'll need a paintbrush to brush soil and sand off plants as you work.
8. With a glue gun, cover a tiny, upside down, 2-inch terra-cotta pot with red fabric.
9. Cut the cork to about ¼ inch and use the glue gun to attach to the clear marble to make the crystal ball.
10. Attach the crystal ball to the table with a glue gun.
11. Add colored beach glass, crystals, or amethyst between plants.
12. Add some puffs of preserved moss to add some instant color.

Plants

Any houseplant store will have a good selection of terrarium-suitable plants. Ask a plant store professional for recommendations. You can find collections of different terrarium plants available online. Some suggested plants:

- *Echeveria pulidonis* (Pulido's echeveria)
- *Myrtillocactus geometrizans* 'Fukurokuryuzinboku' (blue breast cactus)
- *Delosperma echinatum* (pickle plant)
- *Lithops* (living stones)
- *Haworthia fasciata* 'Big Band' (zebra plant)
- *Echeveria* (there are many kinds, *Echeveria runyonii* 'Topsy Turvy' is a great one!)
- *Hatiora salicornioides* (bottle cactus, dancing-bones, drunkard's-dream, or spice cactus)
- *Senecio rowleyanus* (string of pearls)

Care

This project is an open terrarium. In reality, it is many plants sharing one planter.

- **Light:** This terrarium will need bright yet indirect light. North-facing windows are generally best for this type of light. They will perform best with 8 to 10 hours of light per day. A full spectrum grow light will be necessary. Plants grow toward light, so light from above is ideal.
- **Water:** These plants require very little water. After a light initial watering, they may go a month or more without watering. They should never sit in wet soil.
- **Maintenance:** These are slow-growing plants. They will need little maintenance. Always trim dead leaves to keep the terrarium looking its best.

Champagne Cage Chairs

SKILL LEVEL: 1
TIME: 5 MINUTES

What you need

- Champagne cork cage (It's called a muselet!)
- Needle-nose pliers

Instructions

1. Untwist the wire along the bottom of the cage and remove it.
2. Straighten out the wire.
3. Find the center of the wire by curving it gently in half.
4. Using the pliers, work the wire to form the back of the chair, keeping enough length on either end to attach to the seat bottom.
5. Attach the back of the chair to the seat by twisting the seatback wire ends to the chair legs with the pliers.

TERRARIUMS 25

POTIONS CLASSROOM TERRARIUM

"There will be no foolish wand-waving or silly incantations in this class. As such, I don't expect many of you to enjoy the subtle science and exact art that is potion making."

—PROFESSOR SEVERUS SNAPE,
HARRY POTTER AND THE SORCERER'S STONE

SKILL LEVEL: 3
TIME: 1 TO 2 HOURS

Harry Potter might not name Potions as his favorite class for his first five years at Hogwarts, as it was taught by the strangely antagonistic Potions Master Severus Snape. But he wouldn't deny that sleeping draughts, Polyjuice Potion, and Veritaserum had a significant impact in his challenges at the school.

Professor Snape's office in *Harry Potter and the Sorcerer's Stone* was lined with 500 potion bottles, each hand-labeled by the graphics department members with lists of ingredients and serial numbers as well as stains and splashes of liquid. These bottles ranged from several inches to several feet in size, with special bottles made by prop maker Pierre Bohanna from test tubes and other uniquely shaped bottles for a "wizardy" look. Snape's office in *Harry Potter and the Chamber of Secrets* and his fully stocked storeroom in *Harry Potter and the Goblet of Fire* added even more bottles, filled with dried herbs and other plants, small plastic toy animals, and baked animal bones from a butcher shop. Professor Horace Slughorn became Potions Master in *Harry Potter and the Half-Blood Prince*, where the number of potions bottles topped 1,000.

This tall terrarium in a cork-topped jar, inspired by the hundreds of bottles and jars created for the films, will showcase a distinct collection of small tubes and flasks surrounded by a mixture of beautiful plants.

What you need

- Terrarium jar with airtight lid (project uses a jar 8 inches wide and 17 inches tall)
- Lava rock
- Sand
- Scissors
- Sphagnum moss or optional nylon or metal mesh screen
- Horticultural charcoal
- Potting soil
- Preserved moss
- Hammer
- Sandwich baggie
- Test tube with cork top
- Small glass flask with cork top
- Terrarium plant tongs, shovel, and brush (or paintbrush)
- Plants (see Plants on page 30)

Instructions

1. Add a layer of lava rock and sand to the jar.
2. Add a ½-inch layer of sphagnum moss, or a screen cut to the jar's shape, on top of the lava rocks. This will keep soil from mixing with the lava rocks which will form a layer for water to sit in if the terrarium is overwatered, keeping roots out of constant wetness.
3. Place the horticultural charcoal on top of the screen or moss. It doesn't have to be an entire layer. The charcoal will help with moisture retention and mildew control. A scattering should be fine.
4. Add a layer of the potting soil making a "hill"—deeper on one side than the other. This will make the view from the front of the terrarium more layered and allow more room for plants.
5. Crush lava rock inside a sandwich baggie with a hammer and add a layer to both the test tube and the small flask, followed by a few pieces of horticultural charcoal, and then soil. Plant plants in each. Add some preserved moss for color.
6. Place the test tube and flask into the terrarium.
7. Add plants around the test tube and flask. Keep taller plants toward the back of the setting. The back will depend on which direction most will view the terrarium. This is a deep terrarium, you'll need to work with terrarium plant tongs and a shovel to place plants, and a brush to wipe soil off the glass before closing up.
8. Add some puffs of preserved moss to add some instant color. The preserved moss will also help retain moisture for the rest of the live plants.
9. Water the terrarium and close by putting on the cork top.

TERRARIUMS 29

Plants

Any houseplant store will have a good selection of terrarium-suitable plants. Ask a plant store professional for recommendations. You can find collections of different terrarium plants available online. Some suggested plants:

- *Asplenium nidus* (crispy wave fern, bird's nest fern)

- *Cryptanthus bivittatus* 'Pink Starlite' (pink starlite earth star, pink starlite starfish plant)

- *Saxifraga stolonifera* (strawberry begonia)

- *Fittonia albivenis* (nerve plant)

- *Nephrolepis exaltata* 'Fluffy Ruffles' (sword fern, dwarf Boston fern)

- *Thuidium delicatulum* (common fern moss)

Care

This project is a closed terrarium. It is an enclosed ecosystem.

- **Light:** This terrarium will need bright yet indirect light. North-facing windows are generally best for this type of light. They will perform best with 8 to 10 hours of light per day. A full spectrum grow light may be necessary. Plants grow toward light, so light from above is ideal.

- **Water:** After an initial watering and with the top sealed tightly, it may never need watering again. If there is condensation on the inside throughout the day, it will need to dry out by taking the lid off for some time. Ideally, it should possibly have some condensation appear in the morning or evening, but not during the day.

- **Maintenance:** As plants grow, they need to be trimmed and even replaced if they get too large or crowd out other plants. Always trim dead leaves to keep the terrarium looking its best.

MAD-EYE MOODY'S FLASK TERRARIUM

"What's that he's drinking, do you suppose?"

"I don't know, but I don't think it's Pumpkin Juice."

—SEAMUS FINNIGAN AND HARRY POTTER,
HARRY POTTER AND THE GOBLET OF FIRE

SKILL LEVEL: 3
TIME: 1 TO 2 HOURS

The thing about the fourth years' Defense Against the Dark Arts Professor Alastor "Mad-Eye" Moody in *Harry Potter and the Goblet of Fire* was that he really *wasn't* Alastor "Mad-Eye" Moody, a Dark wizard catcher known as an Auror. The real Moody was kidnapped and replaced by Bartemius Crouch, Jr., a Death Eater who escaped Azkaban prison and conspired to bring Harry Potter to a treacherous meeting with Voldemort. Barty Crouch, Jr. used Polyjuice Potion to affect his disguise, which needed to be taken frequently to keep up the ruse, so he kept the potion in a small hip flask.

Mad-Eye's "mad eye" was not a digital effect, but a silicone prosthetic encased in a brass porthole-shaped frame. The pupil inside had a very small magnet attached that was radio-controlled to move around. Occasionally, the magnet came too close to the end of the frame, causing it to lose its link and pop off!

This flask-shaped terrarium gives tribute to the real and transformed Mad-Eye's mad eye as seen in the films.

What you need

- Flask-shaped glass jar with cork stopper
- Lava rock
- Sand
- Scissors
- Sphagnum moss or optional nylon or metal mesh screen
- Horticultural charcoal
- Potting soil
- Terrarium plant tongs, shovel, and brush (or paintbrush)
- Plants (see Plants on page 36)
- 1½-inch bottle cap
- Drill
- ⅞-inch drill bit
- 1/16-inch drill bit
- Screwdriver
- Small screw
- Small washer
- Wire cutters
- Picture-hanging nail
- Glue gun
- Paintbrush
- Metallic gold acrylic paint
- Clear glass marble
- White acrylic paint
- Blue acrylic paint
- Black acrylic paint
- Tape measure
- Leather belt strap
- Pen
- Hobby knife

Instructions

1. Add a layer of lava rock and sand to the jar to cover the bottom.
2. Add a ½-inch layer of sphagnum moss or screen cut to fit inside the jar.
3. Place the horticultural charcoal on top of the moss. It doesn't have to be an entire layer. The charcoal will help with moisture retention and mildew control. A scattering should be fine.
4. Add a layer of potting soil.
5. Plant the plants in the soil using the terrarium plant tongs and shovel. It is a narrow glass jar, so three plants may be enough. Brush off inside glass with the terrarium brush.
6. Add narrow strips of moss between plants.
7. Seal the terrarium with the cork stopper.
8. Drill a hole into the center of the bottle cap with the ⅞-inch drill bit.
9. With a 1/16-inch drill bit, drill a hole on the side of the bottle cap. Screw in a small screw with washer to look like an adjustable "gear" on the side of the bottle cap.
10. With a 1/16-inch drill bit, drill a hole next to the screw and washer "gear." With wire cutters, cut down a picture hanging nail, the type with a notched edge, so that it can be inserted into the bottle cap. Secure with the hot glue gun.
11. Use the paintbrush to paint bottle top, screw, and nail with gold paint. Let dry, paint again.
12. Place the clear glass marble into the inside of the bottle cap, peeking out through the opening. Hot glue in place.
13. Paint the inside of the bottle cap white.
14. Paint a blue circle "iris" onto the marble on the outside of the bottle cap. Let dry.
15. Paint a black "pupil" on top of the blue iris. Let dry.
16. Place the bottle cap "eye" on the leather belt material. Outline the bottle cap with a pen.
17. On the bottle cap outline on the leather, mark off two ½-inch "arms" that will provide an area for gluing the longer leather strips. They should be angled at a 2 o'clock and 10 o'clock position on a clock. Cut out this shape.
18. With the hobby knife, cut leather to create one long ⅜-inch-wide strip. It is important these strips not be too wide that they block light from getting to the terrarium.
19. Cut with a hobby knife two straps long enough to reach from the bottle cap "eye" to meet on the back of the flask. For the flask shown, one was 4 inches, and the other was 6 inches.
20. Hot glue the bottle cap "eye" to the round leather piece with the two "arms."
21. Hot glue the narrow leather strips to the leather base of the eye at the "arms."
22. Wrap the leather straps around the flask. Where they meet, cut with scissors and glue with the hot glue gun.
23. Add a dab of hot glue under the leather eye patch onto the glass flask to keep the leather strap in place.

Plants

Any houseplant store will have a good selection of terrarium-suitable plants. Ask a plant store professional for recommendations. You can find collections of different terrarium plants available online. Some suggested plants:

- *Ficus pumila* 'Quercifolia' (string of frogs, miniature oakleaf fig)
- *Thuidium delicatulum* (common fern moss)
- *Pilea microphylla* (artillery fern)
- *Pellaea rotundifolia* (button fern)

Care

This project is a closed terrarium. It will create and thrive in its own ecosystem.

- **Light:** This terrarium will need bright yet indirect light. North-facing windows are generally best for this type of light. They will perform best with 8 to 10 hours of light per day. A full spectrum grow light may be necessary. Plants grow toward light, so light from above is ideal.

- **Water:** This is an enclosed ecosystem. After an initial watering and with the top sealed tightly, it may never need watering again. Ideally, it should possibly have some condensation appear in the morning or evening, but not during the day. If there is condensation on the inside throughout the day, it will need to dry out by taking the cork lid off for some time.

- **Maintenance:** As plants grow, they need to be trimmed and even replaced if they get too large or crowd out other plants. Always trim dead leaves to keep the terrarium looking its best.

ARGUS FILCH'S LANTERN TERRARIUM

"There was a time detention would find you hanging by your thumbs in the dungeons. God, I miss the screaming."

—ARGUS FILCH,
HARRY POTTER AND THE SORCERER'S STONE

SKILL LEVEL: 3
TIME: 1 TO 2 HOURS

Hogwarts caretaker, Argus Filch, is always on the prowl for misbehaving students. In *Harry Potter and the Sorcerer's Stone*, Harry secretly enters the Restricted Section of the Hogwarts Library to find information on the creator of the Sorcerer's Stone, Nicolas Flamel, but the first book he opens screams at him. This alerts the ever vigilant Filch, who has been stalking the dark corridors, lantern held high to capture his quarry. This time, Harry is able to escape, and Filch fails to find him. Filch uses his lantern again when he accompanies Harry, Hermione Granger, Ron Weasley, and Draco Malfoy after they're caught outside at night and so must serve detention in the Dark Forest, where they will help Hagrid search for a wounded unicorn.

Whenever an actor used a handheld lantern in the early films, they needed to wear large, heavy batteries to power it, hidden in robes and jackets. This lantern, featuring fairy lights, is much less cumbersome and would be much easier to search for midnight troublemakers than the one used in the films!

What you need

- Tall cylindrical glass jar that fits inside lantern
- Lava rock
- Sand
- Scissors
- Sphagnum moss or optional nylon or metal mesh screen
- Horticultural charcoal
- Potting soil
- Terrarium plant tongs, shovel, and brush (or a paintbrush)
- Plants (see Plants on page 42)
- White stone or shell
- Fairy lights
- Clear tape
- Silicone sealant
- Lantern

Instructions

1. Add a layer of lava rock and sand to the jar.
2. Add a ½-inch layer of sphagnum moss or cut a circular screen to the jar's shape and place on top of the lava rocks. This will keep soil from mixing with the lava rocks which will form a layer for water to sit in if the terrarium is overwatered, keeping roots out of constant wetness.
3. Place the horticultural charcoal on top of the screen or moss. It doesn't have to be an entire layer. The charcoal will help with moisture retention and mildew control. A scattering should be fine.
4. Add a layer of the potting soil.
5. Using the terrarium plant tongs and shovel, plant the plants in the soil. It is a narrow glass jar, so two plants may be enough. Plants with white details are used for this lantern.
6. Clean off the glass with the long-handled terrarium brush.
7. Add a white stone or shell.
8. Wrap fairy lights around the outside of the jar: Start with a small piece of clear tape and tape the end of the fairy lights very near the bottom of the jar. Wrap around going up the jar until the top. Add some tape at the top to hold the fairy lights in place.
9. Add a dab of silicone sealant to the inside bottom of the lantern.
10. Place the cylindrical jar inside the lantern, making sure to center it. The silicone sealant will keep the glass jar in place, without having to glue it permanently.
11. With the lantern lid closed, the fairy light battery pack will not be seen. Switch it on and close the lid.

Plants

Any houseplant store will have a good selection of terrarium-suitable plants. Ask a plant store professional for recommendations. You can find collections of different terrarium plants available online. Some suggested plants:

- *Asplenium nidus* (crispy wave fern, bird's nest fern)
- *Cryptanthus bivittatus* 'Pink Starlite' (pink starlite earth star, pink starlite starfish plant)
- *Saxifraga stolonifera* (strawberry begonia)
- *Fittonia albivenis* (nerve plant)
- *Nephrolepis exaltata* 'Fluffy Ruffles' (sword fern, dwarf Boston fern)
- *Thuidium delicatulum* (common fern moss)

Care

This project is an open terrarium. Technically, it is plants sharing a large, shallow planter.

- *Light:* This terrarium will need bright yet indirect light. North-facing windows are generally best for this type of light. They will perform best with 8 to 10 hours of light per day. A full spectrum grow light may be necessary. Plants grow toward light, so light from above is ideal.

- *Water:* Water lightly once a week or when plants show signs of droopy leaves. Watering with a mister (spray bottle) is best. Terrarium plants do like humidity, so a mist between waterings will make them happy. If leaves seem to be turning brown and getting crispy edges, they need more humidity.

- *Maintenance:* As plants grow, they need to be trimmed and even replaced if they get too large, or crowd out other plants. Always trim dead leaves to keep the terrarium looking its best.

CARNIVOROUS PLANT BOG GARDEN

"Are those Tentacula leaves, sir? They're very valuable, aren't they?"

"Ten galleons a leaf to the right buyer—not that I'm familiar with any such back-alley transaction, but one does hear rumors. My own interests are purely academic, of course."

—HARRY POTTER AND HORACE SLUGHORN,
HARRY POTTER AND THE HALF-BLOOD PRINCE

SKILL LEVEL: 3
TIME: 1 TO 2 HOURS

Carnivorous plants won't snap at you like the Venomous Tentacula or encircle you like Devil's Snare—they just happen to like insects or other tiny types of protein as a source of nutrition and rarely if ever grow to a man-eating size. In addition to being a fascinating form of flora, these plants are typically colorful and exotic looking, with leaves like cups or sticky hairs that trap their prey.

The Venomous Tentacula seen in the greenhouse in *Harry Potter and the Half-Blood Prince* is not carnivorous, but it does have tentacles as in *Tentacula*, which contain venom as in *Venomous*. Though writhing around and nipping at him, Potions Professor Horace Slughorn feels this plant is worth the effort to snip off some leaves. There was no danger to the Potions Master, however, as the moving tendrils of the Venomous Tentacula were computer generated in a scene that was unique to the films.

With this Wardian case of carnivorous plants, inspired by the encounters Harry has with the threatening plants in the films, you can impress your friends and frighten your foes—for purely academic purposes of course, as Slughorn tells Harry.

What you need

- Glass case
- Caulk gun
- Silicone caulk
- Lava rocks
- Sand
- Scissors
- Nylon or metal mesh screen
- Horticultural charcoal
- Sphagnum moss
- Soil mix (mostly perlite and peat moss)
- Horticultural grit
- Terrarium plant tongs, shovel, and brush (or paintbrush)
- Plants (see Plants on page 48)
- Moss
- Distilled water, rainwater, or reverse osmosis water

Instructions

1. Use the caulk gun to seal edges of the bottom of the glass case with silicone caulk. Let dry. (Most silicone caulk dries within an hour. Follow caulk instructions.) Test seal by adding water. If it leaks, add more silicone in leaking spots and let dry.

2. Add a layer of lava rock and sand to the case.

3. Use the scissors to cut a screen to the jar's shape and place on top of the lava rocks. This will keep soil from mixing with the lava rocks which will form a layer for water to sit in if the terrarium is overwatered, keeping roots out of constant wetness.

4. Place some horticultural charcoal on top of the screen or moss. It doesn't have to be an entire layer. The charcoal will help with moisture retention and mildew control. A scattering should be fine.

5. Cut some sphagnum moss into smaller pieces, the size of confetti, and mix with potting soil. About ⅓ sphagnum moss and ⅔ soil should be fine for the mix. You'll need about 2 cups of soil for a container this size and about 1 cup sphagnum moss.

6. Add ½ cup of horticultural grit and mix.

7. Add a 1-inch layer of this soil over the charcoal and screen.

8. Plant the plants in the soil, using soil from the pots they came in as much as you can. Build up the soil so the plants are on a hill in the center of the terrarium. Using the terrarium plant tongs and shovel will help in hard-to-reach places. Clean off the glass with the terrarium brush when finished.

9. Add live moss wherever there is bare soil. Break it up and piece it around the plants.

10. Give it a good watering with the distilled water, rainwater, or reverse osmosis water.

Plants

Any houseplant store will have a good selection of terrarium-suitable carnivorous plants. Ask a plant store professional for recommendations. Garden centers that specialize in water plants may have a good selection. You can find collections of different carnivorous plants available online. Some suggested plants used in this terrarium:

- *Utricularia dichotoma* (bladderwort)
- *Sarracenia purpurea* (purple pitcher plant)
- *Pinguicula primuliflora* (pings, southern butterwort or primrose butterwort)
- *Dionaea muscipula* (Venus flytrap)
- *Drosera* (sundew)

Wardian Cases

A Wardian case was the earliest form of terrarium. It was designed in the 1800s by Nathaniel Bagshaw Ward, a London physician. It helped European botanists and plant collectors transport and keep plants found in other parts of the world. Previously, there was no way to ship live plants over long sea voyages. Wardian cases of ferns ended up featured in many European Victorian homes, as well as upper class American homes along the east coast.

Care

Carnivorous plants are NOT like other houseplants. They have sticky hairs, attractive tubes, spiky traps, and mucous-y leaves. They are bog plants and have special planting and care needs. They like low-nutrient soil and boggy conditions.

- *Open or closed terrarium:* These plants can live in a closed terrarium setup and will do fine because they'll like the humidity. But if you want the carnivore to come out in them, an open top or side will allow the occasional bug to cross their paths. Insects are nutritious.

- *Venus flytraps* go dormant in winter. Given that these are indoors and will need much light, they will die off eventually if they don't go dormant annually. Venus flytraps require a cold winter dormancy November through February. To mimic the conditions of their natural habitat, you will need to move them somewhere colder. As days shorten and temperatures drop, the leaves will turn black and the plant will die back to the rhizome. This is normal. You can safely trim off any dead growth. In spring, Venus flytraps will flower. The flowering exhausts the plant. To keep the plant growing vigorously in summer, trim the flower stalk before it gets to about 2 inches in length.

- *Light:* This terrarium will need bright light and will take direct sunlight for a good portion of the day. They will perform best with 12 hours of light each day. A full spectrum grow light is necessary. Plants grow toward light, so light from above is ideal.

- *Water:* These plants like moist soil—but not soaking wet. Their roots need to breathe. NOTE: Only water with distilled water, rainwater, or reverse osmosis water. Water from a tap contains chlorine, fluoride, and other chemicals that will kill these plants.

- *Maintenance:* As plants grow, they need to be trimmed and even replaced if they get too large, or crowd out other plants. Always trim dead leaves to keep the terrarium looking its best.

Chapter 2
INDOOR DÉCOR

◆

"Potent healing herbs are the focus of this essential reference manual for the budding Witch or Wizard Medic. Even the Home User will find this an invaluable friend when in need of a touch of Medical Magic."

—BACK COVER COPY OF *HEALING AT HOME WITH HERBS* BY PHYLLIDA SPORE

Plants

- *Terrarium plants:* Your local houseplant store will carry a variety of small-leafed plants for terrariums. All you need is a tiny sprig, or small piece with a bit of root.

- *Succulents:* Use small sprigs pulled from succulents that have calloused over (takes about 1 week to callous over).

- *Air plants* (*Tillandsia*): If you use these, no soil is needed, just a small bit of sand. Colored sand will look great!

- *Chia seeds:* Grow your terrarium from scratch!

- *Moss:* Find some native mosses in your own backyard or on a nature walk.

Care

- *Light:* Keep in indirect bright light.

- *Water:* It will only need water if it looks dry. It may never need watering if the cork seal is tight. If moisture forms and is constant, it may have too much water and need to be opened and dried out a bit.

- *Maintenance:* If all goes well, it will need upkeep, like trimming or replacing dead plants.

What you need

- Small bottle with cork top
- Small pieces of crushed activated charcoal
- Cactus soil
- Moss
- Wooden skewer or toothpick
- Tweezers
- Sprig of a small terrarium plant (see Plants on opposite page)
- Glass bead, colorful small stone, or tiny shell
- Eyedropper
- Necklace chain
- Charms
- Needle-nose pliers

NOTE: Small bottles are readily available at craft stores, often in multiples, as these make good gifts and favors for events, and craft parties.

Instructions

1. Place a small piece or two of the activated charcoal at the bottom of the small jar. These will help keep moisture down and algae at bay.
2. Place pinches of cactus soil in the bottle, filling at least ⅓ of the way to the top. Tamp down slightly.
3. Add moss and tamp down slightly.
4. Tap a small hole in the moss/soil with a toothpick or wooden skewer and add a cutting/sprig with tweezers from a terrarium plant with small leaves. Tiny ferns work well.
5. Add a colorful stone, a small shell, or glass bead for decoration and color.
6. Water with the eyedropper. Just a few drops should do.
7. Cork the bottle and add the necklace chain. Add charms to the necklace with the needle-nose pliers.
8. Your necklace is now charmed! If it works right, you should never see a Nargle!

LUNA LOVEGOOD'S LIVING NECKLACE

"What an interesting necklace."

"It's a charm, actually. Keeps away the Nargles."

—HERMIONE GRANGER TO LUNA LOVEGOOD,
HARRY POTTER AND THE ORDER OF THE PHOENIX

SKILL LEVEL: 1
TIME: ABOUT 30 MINUTES

When Ravenclaw Luna Lovegood is introduced by Hermione Granger to Harry, Ron, and Neville as they make their way to Hogwarts for their fifth year in *Harry Potter and the Order of the Phoenix*, Hermione remarks on the unique necklace she's wearing. It's a doubled chain made from blue beads that holds a cork from a Butterbeer bottle. According to Luna, it's a charm to keep away magical creatures called Nargles that can cause trouble. As she's still wearing the necklace in her last year at the school, in *Harry Potter and the Deathly Hallows – Part 2*, and hasn't mentioned Nargles being a problem, one can assume it definitely works.

There are charms to bring luck or wealth, charms for safety, charms for courage, and even charms to make a garden grow. Inspired by Luna's, you can put together your own necklace that allows you to carry a charmed living garden with you every day.

Magical Moment

Winogrand's Wondrous Water Plants was an Herbology book by Selina Sapworthy that focused specifically on the water plants among Sir Winogrand's collection in the African Lakes. This cover was made for the Harry Potter films by MinaLima and the graphics department.

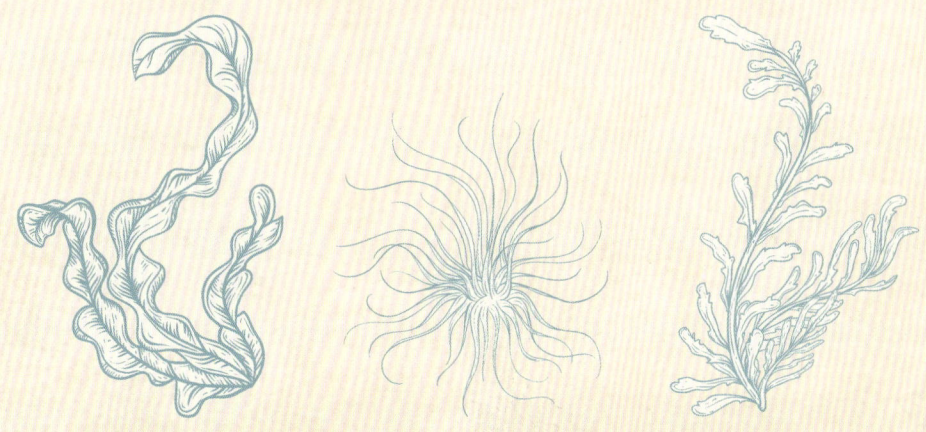

TERRARIUMS 55

Plants

Aquarium stores and most larger pet stores carry a variety of water plants. There's a great variety in leaf shape, color, density, and height. Suggested plants of this project include:

- *Heteroclita* (Asian water fern)
- *Bolbitis heudelotii* (African water fern)
- *Taxiphyllum barbieri* (java moss)
- *Echinodorus amazonicus* (sword plant)

Care

You cannot over water a water plant!

- *Light:* This plant needs bright light, so a full spectrum grow light is necessary. Plants grow toward light, so light from above is ideal.

- *Water:* Water will need to be changed every two weeks or so or when water looks cloudy. NOTE: Any water you add has to be off-gassed by letting tap water sit for 24 hours before using.

- *Maintenance:* As plants grow, they need to be trimmed and even replaced if they get too large. Always trim dead leaves to keep the terrarium looking its best.

TERRARIUMS 53

What you need

- 3 glass jars, and an extra water container
- Terrarium plant tongs
- Water plants (see Plants and Care on page 54)
- Terra-cotta pot (optional)
- Decorative aquarium gravel
- Colander
- Seashells (optional)

Instructions

1. Buy the water plants first. The height and needs of the plants will determine the size of the jar it will fit in. See Plants on page 54 for ideas.
2. The day before you assemble the jars, clean each glass jar inside and out.
3. Fill each jar about ¾ full with tap water. Fill another container with water as well. You'll use this water to top off the jars after plants are placed.
4. Wait 24 hours. The tap water needs to "off-gas" to let chlorine evaporate.
5. Rinse anything that will go into the jars—pots and shells. Rinse the gravel in a colander.
6. Place 1½ inches of decorative aquarium gravel to cover the bottom.
7. Ease a water plant into position, with terrarium plant tongs. Water plants most often need to be settled into a base of gravel to hold them in place.
8. If you're using an optional terra-cotta pot, place it on the decorative aquarium gravel, fill the pot partly with gravel, place the plant in the pot, and add more gravel until the plant is secure.
9. Pour gravel into the jar until the plant's base is covered and the plant is secure.
10. Adjust gravel level to be even.
11. Add optional seashells.
12. Top off the jar with water saved in a separate container.

BLACK LAKE WATER GARDEN

"Now if there's a Tibetan turnip that will allow me to breathe underwater for an hour, then great. Otherwise . . ."

"I don't know about a turnip, but you could always use Gillyweed."

—HARRY POTTER TO NEVILLE LONGBOTTOM,
HARRY POTTER AND THE GOBLET OF FIRE

SKILL LEVEL: 3
TIME: 1 DAY AND 1 HOUR

For the second task of the Triwizard Tournament seen in *Harry Potter and the Goblet of Fire*, Harry must submerge himself under the waters of the Black Lake until he locates and releases a "treasure" taken from him. But how will he be able to breathe long enough to find it? Fortunately, his fellow Gryffindor Neville Longbottom has a great affinity for Herbology—he's read *Magical Water Plants of the Mediterranean*, *Winogrand's Wondrous Water Plants*, and *Sub-Aquatic Botanical Mysteries*, so when he recommends that Harry ingest Gillyweed, he's giving good advice.

It's unknown if Gillyweed itself grows underwater, but there are many plants that thrive in sub-aquatic conditions including varieties of ferns and mosses, which inspired the idea for this dazzling arrangement. Creating your own "under" water garden using jars of different widths and heights, and an eye-catching variety of aquatic plants will make you a winner no matter the task.

SPROUT'S "INTRO TO HERBOLOGY" GARDEN

"Good morning, Professor Sprout!"

—SECOND YEAR HERBOLOGY STUDENTS,
HARRY POTTER AND THE CHAMBER OF SECRETS

SKILL LEVEL: 2
TIME: 2 HOURS

Every aspiring herbologist inspired by Professor Sprout's greenhouse in the films knows it's important to have plants close at hand in their personal greenhouse for healing and well-being and to make breakfasts, lunches, and dinners delicious. These plants are selected to add magic to your meals with fresh herbs growing at your fingertips in a kitchen herb garden you can place on a windowsill or countertop. There is nothing better than fresh herbs such as thyme and dill to cook with to add savory richness to soups, salads, and entrées. Many plants not only taste good in recipes, but they also provide aromas that are inspiring and energizing, such as rosemary or mint. Herbs are easy to grow with adequate sunlight and are a beautiful green addition that will make your kitchen the heart of your house. This box planter will proudly proclaim your interest in and acumen at Herbology.

What you need

- 🥽
- 🎧
- Tape measure
- 1 cedar fence picket, ⅝-by-5½ inches by 6-foot
- Circular saw or handsaw
- Hammer
- Drill and ³⁄₃₂-inch bit
- 12 finishing nails, 1 inch long
- 2-inch tall letters spelling out "HERBOLOGY"
- Flat ½-inch paintbrush
- 2 acrylic paints (green and red were used for this project, choose different ones to make it your own!)
- Pencil
- Wood glue
- 4 terra-cotta pots, 4 to 5 inches wide, with saucers

Instructions

1. From the cedar fence panel, use the circular saw or handsaw to cut a board 24 inches long. Set aside to use later as the front piece.
2. From the cedar fence panel, use the circular saw or handsaw to cut a board 23 inches long. This is the base piece.
3. Cut two 5½-by-5½-inch squares from the cedar fence panel.
4. Assemble by gluing and nailing the 5½-by-5½-inch end pieces to the 23-inch base, keeping all pieces flush, with the ends covering the base ends. Three nails along the bottom of each end should be fine. Pre-drilling the nail hole with a small bit will help the wood to not splinter or split.
5. Turn on its side, glue edges and nail on the 24-inch front piece.
6. Paint your letters. For the project shown, green was used for the front of the letters and edges a rose color. These were interior latex paints on hand. It's always good to go shopping in your own basement or garage for paints! If you buy paint, any acrylic paints will work fine. Letters may need more than one coat.
7. When paint is dry, arrange letters on the front of the box. Make sure to center top to bottom and left to right along the front panel. Draw guidelines lightly in pencil to help keep you on track.
8. Glue each letter down. You may want to use an old paintbrush to spread glue on the backs of the letters uniformly.
9. Once glue dries, add the 4 terra-cotta pots with saucers, filled with your favorite herbs to grow in your kitchen—keep them handy for cooking! See Plants on opposite page or herbs to grow indoors.

Plants

Garden centers and even grocery stores with plant sections will have suitable herb plants for indoor growing. Ask a plant store professional for recommendations or see page 68 for suggestions.

- Basil
- Chervil
- Chives
- Mint
- Oregano
- Rosemary
- Sage
- Thyme

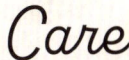

Care

These are plants requiring regular attention.

- *Light:* Direct sun is best. Minimally they need bright sunshine all day indoors. South-facing windows may be best.

- *Water:* These herbs will need water, most likely a couple times a week. Each does have its own requirements. Read tags to know what conditions suit the plants best. Plants with similar requirements grown together will be most rewarding. Other factors may affect watering needs—like direct or indirect sunlight, or drying winds from open windows.

- *Maintenance:* Always trim dead leaves to keep the plants looking their best.

KITCHEN HERBOLOGY

"A Comprehensive Botanical A to Z Directory of Magical Herbs & Fungi For Easy Plant Identification"

—BACK COVER COPY OF *1000 MAGICAL HERBS AND FUNGI* BY PHYLLIDA SPORE

It's clear from Harry's visit to The Burrow for the first time, in *Harry Potter and the Chamber of Secrets*, that the matriarch of the family, Molly Weasley, is the most magical of mothers. She keeps a clean household, with the help of a dishwashing spell, and knits her children warm and wooly clothing, even when she's not holding the knitting needles herself! Harry also notices a stack of cookbooks in her kitchen that include *One Minute Feasts—It's Magic!* and *Enchantment in Baking*. With a notable garden outside the house, it would be no surprise if she brought in fresh herbs to enhance her dishes.

For the film, graphic artists Miraphora Mina and Eduardo Lima contributed their own cookbook idea: *Cooking the Muggle Way*.

It helps to be familiar with herbs, for not only do they lead to great flavor profiles, they can also be beneficial to your health. Following are some of the most popular herbs for growing inside and outside your kitchen.

Basil
OCIMUM BASILICUM

Basil is native to the tropics from Central Africa to Southeast Asia. The French refer to it as "l'herbe royale" ("the royal herb"). Some varieties can grow to more than five feet tall! Its anti-inflammatory properties help to lower the risk of conditions like arthritis, heart disease, and bowel issues. It's one of the easiest herbs to grow!

Chervil
ANTHRISCUS CEREFOLIUM

Also called French parsley, chervil is native to the Middle East. Its name is derived from the Greek word meaning "leaves of joy." Slugs like it too, so you can use it to attract them, if you're into attracting slugs.

Chives
ALLIUM SCHOENOPRASUM

Chives' close relatives include onions, garlic, shallot, leek, and scallions. Chives have insect-repelling properties that many witches and wizards alike use in gardens to control pests. They are native to Europe, Asia, and North America. Their scapes (the stalk and unopened flower buds) and leaves are used for flavors much milder than their onion cousins.

Coriander
CORIANDRUM SATIVUM

Also known as Chinese parsley or *dhania*, all parts of the coriander plant are edible, but most people use the dried seed or leaves. It is native to Southern Europe, Northern Africa, and Southwestern Asia. Coriander is often used in Spanish, Mexican, Latin, and Indian cuisine. It's common in spice rubs, marinades, chilis, sauces, soups, and curries.

Dill
ANETHUM GRAVEOLENS

Dill is in the celery family and was found in the tomb of Egyptian Pharaoh Amenhotep II from around 1400 BCE. Dill makes a great garnish, as it shares its scent with everything else on a plate. It loses flavor the longer it's cooked, so be careful not to overcook it. Fresh dill is a treat in everything from soaps to pickles!

Lavender
LAVANDULA OFFICINALIS

There are 47 known species of lavender, and it is native all over Europe, the Mediterranean, Asia, and India. Its essential oils are used in everything from medicine to cosmetics to cleaning products. It's a strong flavor, so a little goes a long way. It often gets added to sugar, tea, and dry rubs for meats.

Lemon Balm
MELISSA OFFICINALIS

Lemon balm is in the mint family and has naturalized pretty much everywhere but was native to the Mediterranean. The leaves are used as herbs and teas, and in candies and toothpaste. Lemon balm is also used to attract bees for honey production and was cultivated in Thomas Jefferson's Monticello gardens.

Lemon Grass
CYMBOGON CITRATES

Also known as barbed wire grass, silky heads, and fever grass, lemon grass is indigenous to Asia, Africa, and Australia. Its oil is used in soaps and as a natural insect repellent. In beekeeping, its oil imitates honeybee pheromones to attract bees to a hive or to a swarm. Most commonly, lemon grass is used minced or pounded in marinades or grilled meats for a sweet, citrusy flavor.

Marjoram
ORIGANUM MAJORANA

Also called sweet marjoram and knotted marjoram, marjoram is indigenous to the Mediterranean and western Asia. It was known to ancient Greeks and Romans as a symbol of happiness. It is often used in marinades for fish, chicken, and lamb.

Mint
MENTHA

There are many varieties of mint—whether shiny, fuzzy, smooth, crinkled, bright green, or variegated—but they all have that distinctive minty scent. You can always tell a member of the mint family by its unique square stem. If you plant mint, always plant in a pot. In the ground, it grows vigorously and will take over surrounding plants. Young leaves have more flavor than older leaves. Cooks usually prefer spearmint for savory dishes and peppermint for desserts.

Oregano
ORIGANUM VULGARE HIRTUM

Also called wild marjoram, oregano is native to the Mediterranean but has naturalized all over the Northern Hemisphere. Oregano is commonly found in Turkish, Greek, Spanish, Italian, Mexican, and French cuisine. It's often used in tomato-based dishes and most foods using olive oil. Oregano remains flavorful when dried. It's a great plant for a potted windowsill garden.

Parsley
PETROSELINUM CRISPUM

Parsley is native to the central and eastern Mediterranean regions. It is widely used in Middle Eastern, Mediterranean, Brazilian, and American cuisine. Parsley has lots of vitamin K, which is linked to healthy bones. It can be used in almost everything and, because of its looks, is a favored garnish worldwide.

Rosemary
SALVIA ROSMARINUS

Rosemary is a shrub with evergreen, needle-like leaves native to the Mediterranean. The first known mention of rosemary was on stone tablets from 5000 BCE! It's often used as a decorative landscape plant, used often in sensory gardens, and is a great flavor addition to stuffing and roast meats.

Sage
SALVIA OFFICINALIS

Sage is an evergreen shrub, with woody stems, grayish leaves, and bluish flowers. It is a member of the mint family and native to the Mediterranean region. Sage has an earthy, savory flavor, which is a nice addition to heavier dishes like sausage, stuffing, cured meats, and creamy pasta. Sage was also an indication of the presence of mythical fairies.

Tarragon
ARTEMISIA DRACUNCULUS

Also known as estragon and mugwort, tarragon is native to Siberia, and it can grow to 4 feet tall! Tarragon has impressive health benefits, including reducing blood sugar, inflammation, and pain, while improving sleep, appetite, and heart health. Tarragon is used in many French dishes. It is often used in salad dressings, with chicken, and in potato salads. Leaves can be used whole or chopped and added toward the end of a recipe because if cooked too long, they turn bitter. Dried tarragon can be used earlier in recipes, as its flavor is not as strong as fresh tarragon.

Thyme
THYMUS VULGARIS

Thyme is indigenous to the Mediterranean—ancient Egyptians used the plant in their embalming rituals! Romans used it to purify rooms, and Europeans in the Middle Ages used it under pillows to aid sleep and ward off nightmares! Thyme is often used in savory dishes like roasted meats, vegetables, and fish, as well as in baking.

INDOOR DÉCOR

PORTKEY PLANTERS

"That isn't just any old manky boot, mate. It's a Portkey."

—FRED AND GEORGE WEASLEY,
HARRY POTTER AND THE GOBLET OF FIRE

SKILL LEVEL: 1
TIME: 1 HOUR

There are many and varied ways of magical travel in the Wizarding World—by broom or Thestral, Knight Bus, Floo powder, and Apparition, among others. A Portkey is an inanimate object that has been enchanted to send the user to a predetermined location at a predetermined time. It could be anything: a book, a bucket, or a hairbrush. In *Harry Potter and the Goblet of Fire*, Arthur, Fred, George, and Ron Weasley, Amos and Cedric Diggory, Hermione, and Harry grasp a boot to send them to the 422nd Quidditch World Cup.

For the "manky old boot" used by the Weasleys, which inspired this inventive planter, the prop makers bought new footwear, then members of the team wore it around the studio for weeks to "break it down," wanting it to look as if it had been laying in a field for a long time.

As Portkeys can use any object laying around—hopefully at home, and not in a field—any everyday item can be used to disguise a Portkey as a planter that will fool any Muggle.

What you need

- A boot, teapot, or cauldron (or any everyday item that can hold a plant to disguise any Portkey as a planter), drainage a plus
- Lava rock
- Sand
- Stones (optional)
- Scissors
- Sphagnum moss or optional nylon or metal screen
- Horticultural charcoal
- Potting soil
- Plants (see Plants on opposite page)

Instructions

1. The perfect Portkey planter will have drainage holes in the bottom, but if not, judicious and careful watering is fine. Look for items around your house, basement, and attic. A second-hand store is a good place to find the perfect Portkey.

2. Add a layer of lava rock and sand to the bottom of the Portkey planter, an inch should do. In the case of the boot, or a similar planter, the plant sits inside a planter covered in a plastic bag to catch any drained water. Weigh the toe of the boot down with stones so it won't topple, if needed.

3. Add a ½-inch layer of sphagnum moss or a piece of cut screen, on top of the lava rocks. This will keep soil from mixing with the lava rocks which will form a layer for water to sit in if the planter is overwatered.

4. Place a layer of horticultural charcoal on top of the screen or moss.

5. Add potting soil 1 to 2 inches from top of planter.

6. Add a plant to the portkey. One good-sized plant per portkey is fine.

Plants

Any houseplant store, garden center, or big box hardware store will have suitable houseplants. Ask a plant store professional for recommendations! The plants used in the portkey planters shown:

- *Nephrolepsis biserratta 'macho'* (macho fern)
- *Platycerium bifurcatum* (Elkhorn fern)
- *Chlorophytum comosum* (spider plant)

Care

These are normal houseplants requiring regular attention. If using the above mentioned plants, care is as follows:

- *Light:* Bright yet indirect light is generally best. Most houseplants are understory tropical shade plants requiring dappled light or light shade. North-facing windows are generally best for this type of light. They will perform best with a minimum of 8 hours of light per day.

- *Water:* A weekly watering is fine for houseplants such as these. Monitor the soil to make sure it's not too wet (or dry!). Droopy leaves mean it needs water. Leaves with dry or brown edges mean it needs more humidity.

- *Maintenance:* Always trim dead leaves to keep the plants looking their best.

CORNISH PIXIE AIR PLANT CAGE

"Freshly caught Cornish Pixies. Laugh if you will, Mr. Finnigan, but pixies can be devilishly tricky little blighters."

—PROFESSOR GILDEROY LOCKHART,
HARRY POTTER AND THE CHAMBER OF SECRETS

SKILL LEVEL: 4
TIME: 5 TO 6 HOURS

For his opening lesson for second-year students, Defense Against the Dark Arts Professor Gilderoy Lockhart may think that using Cornish Pixies will satisfy his task of teaching them about the foulest creatures in wizardkind, but these blue-skinned mischief makers are laughable compared to the werewolves, trolls, and vampires that fill his popular books. Then again, as he finally confesses, he never actually encountered those beasts. So, it's no surprise he can't control the Cornish Pixies as they escape their cage and wreak havoc in the classroom.

The special effects crew used thin wires to pull at the books and papers that fell or flew during the pixies' destruction, with the actors adding to it by knocking off nearby props. There were twenty digitally created pixies seen in *Harry Potter and the Chamber of Secrets*, placed on screen at different levels in the background, midground, and foreground to create a sense of depth.

Bring a bit of the Wizarding World films into your home with this air plant cage inspired by the Cornish Pixies in the film. The air plants used herein will keep this Cornish Pixie from causing trouble without having to use the charm *Immobulus*.

What you need

- 🖐
- Decorative birdcage
- Aluminum foil
- Florist wire
- Needle-nose pliers
- Wire cutters
- Sewing pins
- Light-weight oven-bake clay
- Burnishing tools for clay with a variety of unique ends
- Two black glass beads
- Oven
- Paintbrushes: flat, round, and rigger brushes
- Ultramarine acrylic paint
- Turquoise acrylic paint
- Light blue acrylic paint
- Pink acrylic paint
- Hobby knife
- Clear plastic (like from a deli container)
- Superglue
- Scissors
- Fishing line
- Aquarium gravel
- Air plants (*Tillandsia*) (see Plants below)

Instructions

1. First, find a decorative birdcage. This will determine the size of the Cornish Pixie you make and size and number of air plants.
2. Create the Cornish Pixie by tightly wadding up aluminum foil to make a torso shape.
3. Use the needle-nose pliers and wire cutters to cut and shape floral wire to create individual arms, legs, and a neck. Stick each into the foil torso.
4. Create a round head shape from aluminum foil. Cut metal screens into the shape of ears and attach to the head piece with sewing pins. Also add floral wire to the top of the head to create two antennas. Do not attach the head to the body yet.

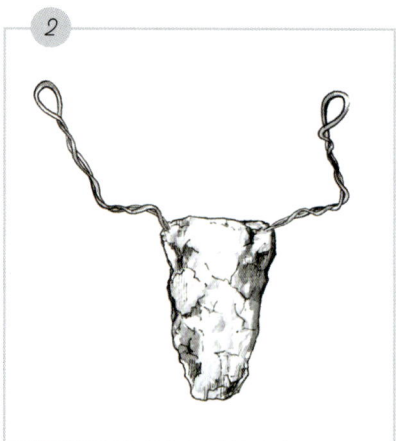

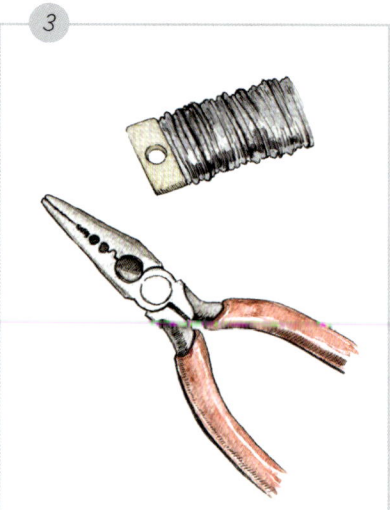

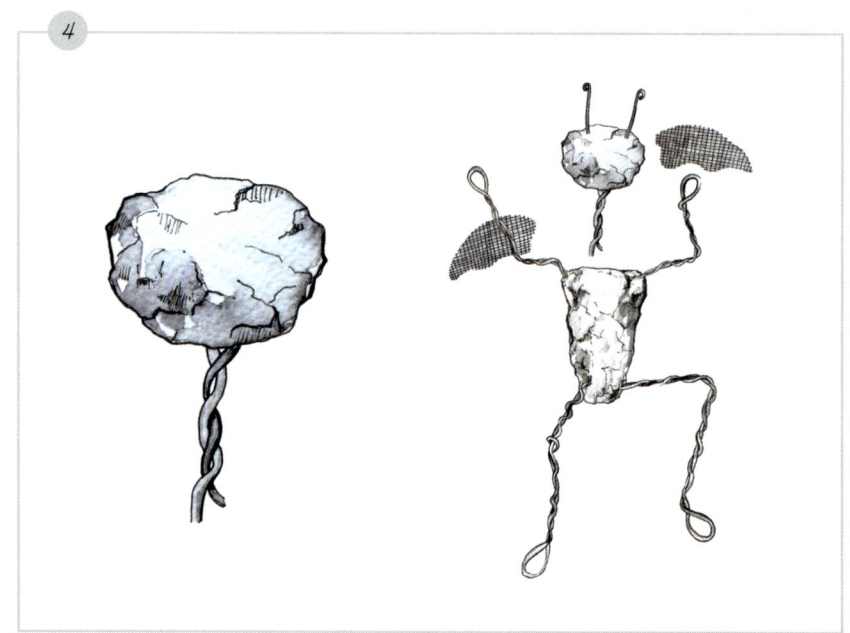

5. Cover the torso with modeling clay and sculpt over the aluminum foil. Bake according to the directions on the modeling clay (usually it's on a glass or metal baking sheet at 275°F for 15 minutes). Let cool.

6. Once cool, add the head to the torso onto the wire metal neck. Cover with clay and sculpt the face and ears. Pinch to make eyes, add clay, and blend to make cheeks, eyebrows, and nose. Slit clay for eye positions. Add the black glass beads. Do not use plastic beads. Plastic beads will melt in the oven.

7. Once you're happy with the face (it's the hardest part, but it doesn't have to be perfect—you're creating a pint-size Cornish Pixie, it doesn't have to be pretty!), bake again. Let dry.

8. Once cool, add clay to create the arms. These are more easily done by flattening out an upper arm's length (the Pixie's arm length, not yours!), folding it over the floral wire arm, and blending the overlap, then repeating for the other arm. Bake and let cool.

9. Once cool, add hands. You may need to add more floral wire to make digits. These Cornish Pixies have three fingers and three toes. Your particular breed can have as many or as few as you wish. Bake and let cool.

10. Once cool, you can start painting: An overall coat of ultramarine blue first will provide a good base coat.

11. Paint the entire Pixie with a turquoise blue.

12. Pixies have whitish highlights. Brush white paint lightly onto the chest, stomach, shoulders, joints, and cheeks. Blend in with a sponge.

13. Paint white details with the fine rigger brush: teeth, claws, and nails if they're visible.

14. Cut out Pixie wings from clear plastic with a hobby knife. Use the template found at www.InsightEditions.com/HarryPotterHerbologyMagic. Scale it up or down for your own Cornish Pixie's size.

15. With superglue, attach wings to the back of the Cornish Pixie.

16. Hang the Cornish Pixie from the top of the cage with fishing line. Tie to the cage top and around his neck to make the Pixie look like it's floating.

17. Add a layer of aquarium gravel to the bottom of the cage. This forms a base for an air plant to sit on, allowing some air circulation under the air plant.

18. Add air plants to the inside of the cage.

19. Add air plants to the outside of the cage by hanging floral wire and setting air plants on the wire(s).

20. Find a bright sunny spot to hang.

INDOOR DÉCOR

Plants

Air plants (*Tillandsia*) are a curious genus of plants, with more than 650 varieties. They are epiphytes—plants not needing soil for nourishment. Most houseplant stores will carry a variety of air plants. Some suggested plants, used in this project:

- *Tillandsia bulbosa* "clump"
- *Tillandsia caput-medusae*
- *Tillandsia xerographica x capitata*
- *Tillandsia xerographica juncifeliam*

Care

This project is an open-air terrarium. They perform best with good air circulation with temperatures between 60°F and 80°F. If you live in an area with cold winters, they would be happy outdoors in a protected spot in summer. Air plants are happiest in a moist environment. If your home is dry, a humidifier will be necessary.

- *Light:* Keep in indirect bright light.
- *Water:* Air plants need to be submerged in water every week or so, as often as you would water any houseplant. A misting every few days with a spray bottle will keep them happy between waterings.

HOGWARTS HOUSES MACRAMÉ PLANT HANGERS

"Now, in a few moments, you will pass through these doors and join your classmates, but before you can take your seats, you must be sorted into your houses. They are Gryffindor, Hufflepuff, Ravenclaw, and Slytherin."

—PROFESSOR MINERVA MCGONAGALL,
HARRY POTTER AND THE SORCERER'S STONE

SKILL LEVEL: 2
TIME: 2 HOURS

The four houses of Hogwarts are named after the school's founders: Godric Gryffindor, Helga Hufflepuff, Rowena Ravenclaw, and Salazar Slytherin. Once sorted by the Sorting Hat, each house reflects the founders' preferences, with different colors and mascots, which inspired these "house" planters based on the films.

Gryffindors are known for their courage, bravery, and determination. Their crest features a lion framed by the house colors of red and gold. Hufflepuff students are known for dedication, patience, and loyalty, with a crest that displays a badger within the house colors of yellow and black. Ravenclaws are known for their wit, learning, and wisdom. House colors of blue and silver showcase their eagle mascot. Slytherins are known for their pride, ambition, and cunning, though any students, including Slytherins, can prove to be as brave and loyal as any Gryffindor. Slytherin's crest shows a snake coiled on a field of its house colors, green and silver.

Whichever house you're sorted into can be honored by a macramé plant hanger that proclaims your house pride. Even better, honor all four with the ancient art of macramé, where cords or strings are knotted in decorative patterns. Crafting these knots takes the same level of skill as tying a shoe.

What you need

- Tape measure
- 4 small terra-cotta pots
- ½-inch paintbrush and a ⅝-inch paintbrush
- White acrylic paint
- Sixteen 1-inch wooden beads
- Tan, black, gray, and golden-yellow acrylic paint
- Clear varnish
- Scissors
- 1 skein each of 3mm macramé cord in dark red, dark green, dark blue, and yellow
- 1 piece driftwood or branch, approximately 4 feet long (or four 2-inch metal rings to hang separately)

Instructions

1. Use the paintbrush to paint the terra-cotta pots white with the exterior house paint. Allow to dry completely.
2. Paint the beads: Paint 4 black, 4 tan, 4 gray, and 4 golden yellow. Allow to dry completely.
3. Paint the top band on each of the 4 pots to match the beads: Paint 1 black, 1 tan, 1 gray, and 1 golden yellow. Allow to dry completely.
4. When dry, add a coat of clear varnish on the beads and the top bands of the pots.
5. Cut 4 strands of red cord, 3 yards each.
6. Fold the 4 strands in half, giving you 8 strands.
7. Mark the four places on the branch where the macramé will hang by dividing the length of the branch into five equal parts and marking them with a pencil.
8. Place the looped strands under the branch on one of the marks. Bring the loop around to the front and pull the cords through the loop to tighten. This is called a lark's head knot.
9. Separate into 4 pairs of cords and measure 12 inches from the branch. Make an overhand knot by making a loop and pass the end through it. Tighten to form the knot.

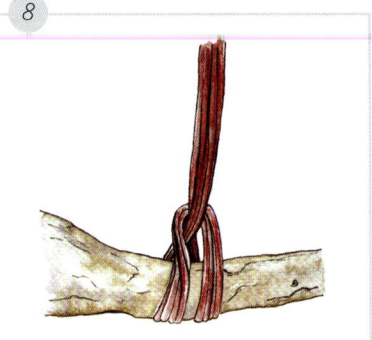

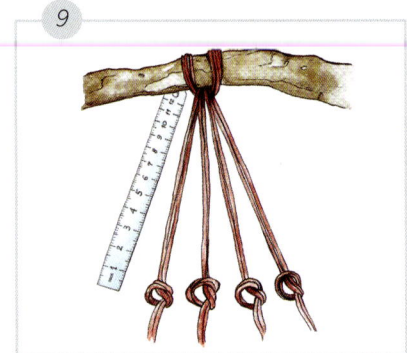

10. Repeat this knot on all 4 sets of cords. Slide a golden yellow bead onto the 4 sets of cords and tie a second overhand knot under each bead to anchor it in place.

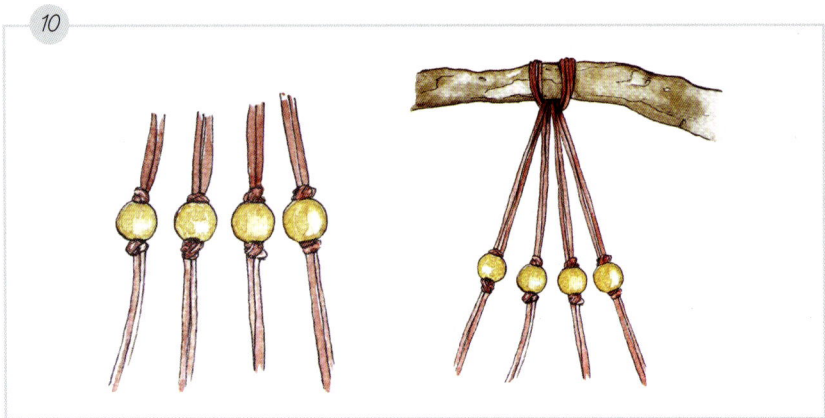

11. Divide the 8 strands into 4 new pairs by matching one strand from each prior pair with a strand from an adjacent pair. Measure down 6 inches and make an overhand knot in each set of cords.

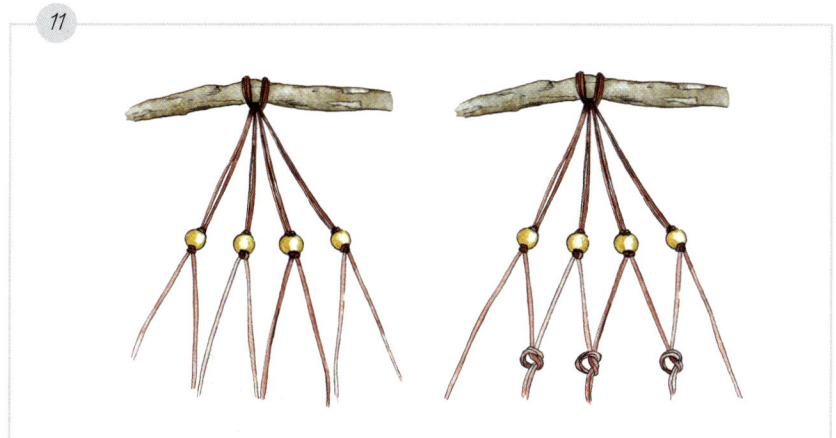

12. Measure 5 inches below the second set of overhand knots, gather all 8 cords together and tie 1 big overhand knot. Pull tight. Cut the cords 6 inches or at desired length.

13. For the second hanger, cut 4 green cords, 2½ yards long each. Fold the 4 strands in half, giving you 8 strands. Measure 8 inches along the branch from the first plant hanger, and place the looped strands under the branch and make another lark's head knot.

14. Separate into 4 pairs of cords and measure 6 inches from the branch. Make an overhand knot in all 4 pairs of cords. Slide a gray bead onto the 4 sets of cords and tie a second overhand knot under each bead to anchor it in place.

15. Repeat steps 5 through 11 to complete your second plant hanger.

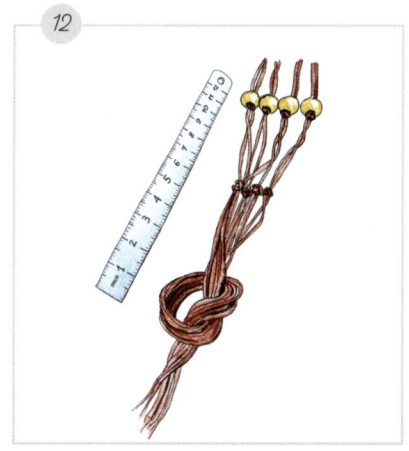

16. Make a second long plant hanger using the blue cord and tan beads and a second short plant hanger using the yellow cord and black beads.

17. Plant the planters with houseplants to use indoors and outside in the summer months. Plant with flowering annuals if the pots are intended for a protected area outdoors.

18. Place the pot that coordinates with the painted band with the painted beads in each plant hanger.

NOTE: Hangers can all be made on the same branch or on 4 separate metal rings, depending on the effect that you want. If you'd rather have 4 individual plant hangers, merely substitute the four 2-inch metal rings for the branch and start each with the lark's head knot and go from there.

INDOOR DÉCOR 83

Plants

Any houseplant store, garden center, or big box hardware store will have suitable houseplants. Ask a plant store professional for recommendations! The plants used in the Hogwarts House color macramé planters shown:

- Ravenclaw: *Senecio cineraria* (dusty miller)
- Hufflepuff: *Petrosedum rupestre* 'Angelina' (creeping sedum)
- Slytherin: *Asplenium nidus* (crispy wave fern)
- Gryffindor: *Kalanchoe blossfeldiana* 'Kalanchoe'

Care

These are normal houseplants requiring regular attention, but care will vary based on what plants you choose. For the plants mentioned above, reference the following.

- *Light:* Bright yet indirect light. Most houseplants are understory tropical shade plants requiring dappled light or light shade. North-facing windows are generally best for this type of light. They will perform best with a minimum of 8 hours of light per day.

- *Water:* A weekly watering is fine for houseplants such as these. Monitor the soil to make sure it's not too wet (or dry!). Droopy leaves mean it needs water. Leaves with dry or brown edges mean it needs more humidity.

- *Maintenance:* Always trim dead leaves to keep the plants looking their best.

INDOOR DÉCOR 85

NEVILLE LONGBOTTOM'S PROPAGATION STATION

"Professor Sprout tells me you have an aptitude for Herbology."

—PROFESSOR ALASTOR MOODY,
HARRY POTTER AND THE GOBLET OF FIRE

SKILL LEVEL: 3
TIME: 1½ HOURS

Propagating plants from cuttings is one of the easiest methods to grow future plants for your own garden or to share with friends. This propagation station is a beautiful way to prepare future generations of your plants in convenient tiers that will allow you to eye your shoots as they grow roots.

Perhaps if Neville Longbottom had had a station such as this, he would have had an easier time with his *Mimbulus mimbletonia* plant in the films. In his earlier years at Hogwarts, Neville may have been considered all fingers and thumbs, but those thumbs were obviously green, as he showed great knowledge and skill when it came to Herbology (disregarding his fainting episode with Mandrakes). In his fifth year, Neville cares for the difficult *Mimbulus mimbletonia* plant with a talent like no other student. In a deleted scene in *Harry Potter and the Order of the Phoenix*, he tends to the prickly plant's needs, but inadvertently pokes the wrong spot and the plant squirts a greenish sludge at him. The scene needed to be filmed several times, with actor Matthew Lewis struggling not to flinch when he knew what was about to happen.

The worst that can happen with *this* propagation station is perhaps getting splattered with a few drops of water.

What you need

- 🥽
- 🎧
- Tape measure
- Three 1½-by-¼-by-15-inch boards
- Handsaw or circular saw
- Small drill bit
- ¾-inch drill bit
- Clamp
- Sandpaper
- Scissors
- Leather cord or twisted jute twine
- 15 test tubes with flared tops
- Optional: paint or stain of your choosing

Instructions

1. If you haven't already, measure out and cut three 15-inch boards. For this project, a long piece of reclaimed lath was used.
2. Measure ½ inch from the ends of each board and drill a hole with a drill bit the size of your leather cord or twisted jute twine.
3. Mark up the holes for the test tubes to slot into by measuring the center hole first (7½ inches in from either end). Mark the center lengthwise (¾ inch from one side of the board).
4. Mark up the spots for the other test tube holes, evenly spaced on each 15-inch board.
5. Stack the three boards on top of each other. Clamp and use the ¾-inch drill bit to drill holes through all three boards at the same time.
6. Reposition the boards and clamps and drill the other holes.
7. Make sure the test tubes fit in the holes you've drilled. You'll likely need sandpaper to open up the holes to accommodate each test tube. The flared test tubes should fit snugly in the holes.

NOTE: The lath board used for this project already has a rustic look and a white paint patina. If you're inclined to paint or stain to coordinate with your home's décor, now's the time to do it.

8. Cut a 44-inch length of the leather cord or twisted jute twine. Find the center at 22 inches.
9. Space the boards 11 inches apart on a table.
10. Feed the cord through the small end holes. Leave slack above the top board to be able to hang the finished project from a nail or hook, about 72 inches.
11. Make a small knot below the top board on both ends, careful to tie it close and tight to the board.
12. The next board should be 11 inches from the top board, allowing for the length of the test tubes hanging down and the plant cutting height.
13. Feed cord through end holes on this second board. Make a small knot below the board, careful to tie it close and tight to the board on both sides.
14. Place the third and final board 11 inches from the second, feed cord through end holes. Make a small knot below the board, careful to tie it close and tight to the board on both sides. Cut off excess cord.
15. Fit the test tubes into each board.
16. Hang, water, and add plant cuttings!

INDOOR DÉCOR 89

Plants

You can take cuttings from most houseplants with leaves. Cut a strip almost the length of the test tube, with leaves sticking out of the test tube. Strip lower leaves off to fit the test tube size. It will take 1 to 2 weeks for most plants to start showing roots. Once there are plenty of roots and the test tube is looking crowded, it's time to move a cutting into a small pot with soil and drainage.

Care

These are normal houseplants requiring regular attention.

- *Light:* Bright yet indirect light. North-facing windows are generally best for this type of light.

- *Water:* Keep water levels just below test tube tops. Depending on how long a cutting takes to root, you'll want to change out the test tube's water every couple weeks or when it appears cloudy.

- *Maintenance:* Always trim dead leaves to keep the plants looking their best.

Propagation Methods

There are many ways to propagate plants. In addition to the ones listed below, there are methods such as root cuttings, layering, grafting, and more. The most common for houseplants are:

By Water
(USED FOR NEVILLE LONGBOTTOM'S PROPAGATION STATION)

Make a cut with clean scissors, or better yet, a precision hobby knife. Scissors tend to squeeze the cut end. You'll have better luck with open cuts. Strip lower leaves off to accommodate the receptacle, make sure the nodes where the leaves were are submerged in the water. Once roots are 1½ to 2 inches long, you can plant it in a small planter with drainage. For the first few weeks you'll want to keep it moist, but not soggy.

By Division

Propagating by division works well with plants that grow offshoots. Simply remove the offshoots (or "pups") with as much of the roots it has as possible. You shouldn't have to cut anything. Repot the offshoots in a small pot with drainage and keep moist for a few weeks until there are signs of growth. Then water like any normal houseplant.

By Leaf

Plants like *Peperomia* and snake plants can be propagated with leaf cuttings. Take a cutting that has some of the stem attached to the leaf and pop it into a small planter with drainage and keep moist for a few weeks until you see some growth. Be patient!

Succulent Propagation

There's a few ways to propagate succulents, depending on the type of plant you have. Succulents, with the right conditions, are the easiest plants to grow, but they like little water and lots of light. Special cactus and succulent soil will help tremendously. Most plant stores, garden centers, and home "big box" stores carry it.

- Healthy, happy succulents will often have plantlets, or offshoots. These can be separated and grown easily in cactus and succulent soil.
- Some grow "pups" on their leaves or stems. These can easily be separated and planted in cactus and succulent soil.
- Others will grow offshoots from their root base. You can gently tease the roots from the parent and pup and plant separately in cactus and succulent soil.
- Cutting requires time for the cut leaves to "heal" by letting them dry out. They'll start growing roots once they are healed. Patience is required!
- Beheading! Some succulents get tall and spindly. Cut off their heads leaving an inch or so of stem. Replant.

WHOMPING WILLOW BONSAI TREE

". . . Not to mention the damage you inflicted on a Whomping Willow that's been on these grounds since before you were born."

—SEVERUS SNAPE,
HARRY POTTER AND THE CHAMBER OF SECRETS

After Harry Potter's been locked in his bedroom at the end of the summer by his uncle, Vernon Dursley, in *Harry Potter and the Chamber of Secrets*, he's rescued by Ron Weasley and his twin brothers, Fred and George, who are driving a flying car. The light blue car is their father's, who cast a spell on it to make it fly. He even included an invisibility booster, so Muggles could not see it. Ron and Harry use the car again when they can't get through to Platform 9¾ and miss their scheduled trip on the Hogwarts Express. The car flies them to Hogwarts and then conks out, crashing into a huge Whomping Willow tree. The name is not inaccurate—the car is thoroughly whomped by the tree's huge branches, and eventually spat out to disappear into the Forbidden Forest.

The creation of the Whomping Willow was a collaboration between the art, visual effects, and special effects departments, which resulted in a hydraulically controlled eighty-five-foot tall tree. First, a base was built, hidden by a rubber-covered trunk that held the car. Then hydraulic branches were added that could grab and surround the car.

Unlike the car in the films, there's no danger of the toy car used here being pummeled—just the fun of an eye-catching showpiece for your garden.

Note: This project is made in two parts, creating an optional handmade hypertufa planter and then planting it up with a bonsai and model car. Hypertufa is a mix of Portland cement, perlite, and peat moss that can be poured into a mold. When dried, it looks like rough stone but is lightweight, porous, and durable.

Hypertufa Planter

SKILL LEVEL: 3
TIME: 24+ HOURS

What you need

- 🧤
- 😷
- Tape measure
- 2 molds: cardboard or plastic containers, with one that fits inside the other with a ¾-inch border between them. For this project, the outer mold is 12-by-7-by-5 inches tall. The inner, smaller mold is 10-by-4½-by-4½ inches tall.
- Non-stick cooking spray
- Rubber gloves
- 2 quarts Portland cement (not quick-set cement)
- 2 quarts perlite
- 2 quarts peat moss
- Large plastic pan for mixing cement, minimum 12 inches by 18 inches by 5 inches deep
- Concrete trowel
- Concrete dye (optional)
- Weights, such as sand or rocks (optional)
- Plastic, like a garbage bag
- Drill with masonry bit

Instructions

1. Put on your gloves!
2. Prepare the molds: Spray the inside of the larger mold and the outside of the smaller with the non-stick cooking spray.
3. Mix the Portland cement, perlite, and peat moss in the large plastic pan. Add water to the mixture until it's thick enough to pick up by hand, about the consistency of cottage cheese. This is your hypertufa mix.
4. Mix in the optional concrete dye into the hypertufa mix now.
5. Scoop hypertufa mix into the larger mold to about 1 inch thick on the bottom, add the smaller container and build up the mixture along the sides. You may have to put a weight inside the smaller mold to hold it down within the larger mold.
6. Cover the molds loosely with plastic; it helps the hypertufa to dry slowly, making it stronger.
7. After 24 hours, remove the plastic and separate the molds.
8. When completely dry, with a masonry bit, drill a drainage hole in the bottom of the larger mold. Don't skip this step. Plants like good drainage!
9. Plant a Whomping Willow (page 96) or other plant inside.

Whomping Willow Planting

> SKILL LEVEL: 1
> TIME: 1 HOUR

What you need

- Hypertufa planter (see opposite page) (or other planter, as long as it is 4 to 5 inches deep and has drainage)
- Small piece of nylon or metal mesh screen or landscape fabric large enough to cover the drainage hole
- Potting soil
- Ponytail palm (see Plants on page 96) or other miniature tree
- Preserved moss (optional)
- Light blue toy car
- Planter saucer

Instructions

1. Put the screen or landscape fabric over the drainage hole inside the hypertufa planter. It's there to keep soil from running out the drainage hole.
2. Add a couple inches of potting soil over the bottom of the planter.
3. Transplant the miniature tree, adding soil as needed to make sure to keep roots as buried as they were in its original container. Keep adding soil until it's about 1 inch from the top of the planter.
4. Optional: Preserved moss can be added to the soil surface for quick effect.
5. Add your light blue flying car (Ford Anglia 105E was used in *Harry Potter and the Chamber of Secrets*). The car shown was found in a crafts store.
6. If you're using the hypertufa planter, or another planter *without* a built-in water reservoir base, you'll need to place the planter on something to catch the water that drains, such as a planter saucer.

Plants

In this project, a ponytail palm was used. You could also use quick-growing weeping, Australian, or dragon claw dwarf willows.

- Ponytail palm, also called elephant foot (*Beaucarnea recurvata,* Zones 9 to 11), is a slow grower and eventually (after many years) may need to be replanted in a larger pot. It does like confined space. Technically, it is neither a tree nor a palm—it is a succulent in the agave family! They are readily available in houseplant stores, garden centers, and even "big box" home stores.

Care

- *Light:* Keep in indirect, bright light. They will perform best with 8 to 10 hours of light per day.

- *Water:* Keep soil fairly dry. When you do water, soak thoroughly allowing water to drain through hole in bottom of planter. Empty planter saucer of water. The planter should not sit in water. Water less in winter months.

- *Maintenance:* As the plant grows, it may need to be replaced if it gets too large for the container. It will be happy outdoors, out of direct sun in summer. It will need to come indoors in areas with cold winters.

True Bonsai Trees

A bonsai, "tray planting," (pronounced *bone-sigh*) is the Japanese art of training miniature trees in shallow pots, making small trees mimic their larger counterparts. Bonsai was developed from the traditional Chinese art of *penjing*. Bonsai can be created from nearly any woody-stemmed tree or shrub and can be cultivated to remain small by crown and root pruning. Some trees are often used as bonsai because they have small leaves or needles, like Japanese maples, boxwoods, firs, figs, and hornbeams, among others. A bonsai tree would work for the Whomping Willow, but the art of bonsai takes much care, skill, special tools, planters, training, and patience to be successful.

Chapter 3

OUTDOOR DÉCOR

◆

*"Pull yourself together, man. You're going into the forest, after all.
Got to have your wits about you."*

—ARGUS FILCH,
HARRY POTTER AND THE SORCERER'S STONE

PROFESSOR FLITWICK'S CHARMED EVER-RAINING RAIN CHAIN

"Swish and Flick!"

—FILIUS FLITWICK,
HARRY POTTER AND THE SORCERER'S STONE

SKILL LEVEL: 4
TIME: 6 TO 8 HOURS

Professor Filius Flitwick teaches Charms, a core subject, at Hogwarts; his first lesson is the charm *Wingardium Leviosa*, which levitates a feather. Other charms students learn at Hogwarts include *Accio*, to bring something to the caster; *Expelliarmus*, to disarm an opponent; and *Oculus Reparo*, which Hermione uses frequently to fix Harry's eyeglasses.

In addition to charms and spells that came from the Harry Potter books, there were several created specifically for *Harry Potter and the Deathly Hallows – Part 2*, as the professors at school raise a shield to protect the school. Used in combination, they are *Protego Maxima* to shield; *Repello Inimicum*, to drive away all enemies; and *Fianto Duri*, which strengthens the two other charms.

One spell would work perfectly when setting off this recirculating rain chain made from miniature terra-cotta pots, so impress your friends by casting *Meteolojinx* when you want to make the water flow.

What you need

- 🥽
- 🧤
- Tape measure
- Wire cutters
- 15 feet of 10-gauge copper wire
- Needle-nose pliers
- Vise
- 1-foot piece of ¾-inch round pipe
- Thirty 2-inch terra-cotta pots (makes a chain about 90 inches long)
- 5- to 6-inch metal rod
- Shovel
- 10-gallon bucket or pail
- Submersible water fountain pump with at least a 10-foot lift, with a flow rate of around 3,500 liters/hour
- 10-foot hose
- Heavy-duty shutoff garden hose water valve
- Heavy-duty wire mesh, about 3-foot round
- Decorative river rocks

Instructions

1. Determine where the rain chain will hang. The best spot will be under a rain gutter that allows a large bucket to be buried directly beneath the downspout of the rain gutter. You will also need an electrical outlet nearby to plug in the fountain.
2. With the tape measure, measure the distance from gutter downspout connection to the ground.
3. Set up a work area where you can hang the rain chain while working on it. It should be the length you measured from downspout to ground.
4. Use wire cutters to cut 35 pieces of 10-gauge copper wire into pieces 14 inches in length.
5. Loop one end of the wire to form a hook. Do this for all 35 pieces.
6. Take the pipe and place one end tightly in a vise.
7. Wrap the copper wire pieces around the pipe tightly, leaving some straight slack on the un-looped end.
8. Place inside one terra-cotta pot.
9. Squeeze the coil to fit inside the pot, with the straight slack sticking through the bottom of the terra-cotta pot.
10. With needle-nose pliers, make a hook in the straight slack sticking out the bottom of the pot.
11. With needle-nose pliers, bend the other hooked end on the top so that it is centered above the pot.

5-11

12. Repeat this with all wires. Instructions are for a 90-inch length which will be around 30 tiny pots. You may need more, or less, depending on the length for your particular rain chain set up.

13. Hook each pot to the next as you work.

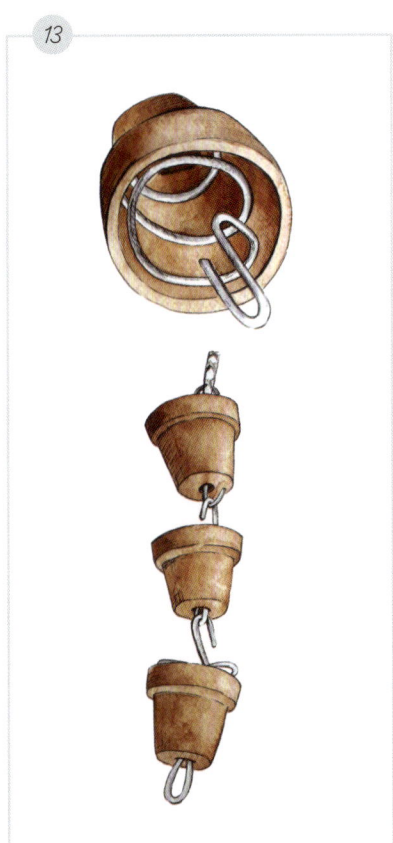

14. Take the rain chain to its final destination and hang it. Put the metal rod inside the downspout in the gutter over the opening. Attach the top rain chain hook to the metal rod through the downspout opening.
15. Where the rain chain hangs near the ground is where you have to bury the 10-gallon bucket or pail as close to the center under the rain chain as possible.
16. Dig a hole large enough to fit the pail inside, keeping the top of the pail at ground level.
17. Place the submersible pump in the pail.
18. Attach one end of the hose to the pump following the instructions that come with the pump. Run the hose up to the rain gutter. To make the illusion work best, it's best to hide or disguise the hose so it's not seen by Muggles.
19. Where the hose ends in the rain gutter, add the heavy-duty shutoff garden hose water valve. This will allow you to control the water flow through the rain chain.
20. Test the system by filling the bucket with water and plugging in the pump. As long as the pump is the right strength, the water will flow. Control the amount of flow by adjusting the shutoff valve.
21. Once it's working to your satisfaction, add the heavy duty wire mesh over the top of the bucket.
22. Place decorative river rocks over the mesh. Hide or cover the hose and electrical cord from the bucket.

Care

If the rain chain is near a tree, the terra-cotta pots may collect occasional debris. Keep them free of debris to run smoothly. If you live where winters freeze the ground, the pump will have to come out and be stored until spring. The pail should be covered to not collect snow and water. Water in the pail may freeze and cause it to crack.

WIZARDING WAYFINDING SIGNPOST

"Welcome to the charming village of Budleigh Babberton, Harry."

—ALBUS DUMBLEDORE,
HARRY POTTER AND THE HALF-BLOOD PRINCE

SKILL LEVEL: 4
TIME: 8 TO 10 HOURS

Throughout the course of his adventures, Harry Potter finds himself in previously unseen magical locations that live side-by-side with the Muggle world. Diagon Alley is a long and winding alley where wizards shop for brooms and robes, bank, and enjoy a Butterbeer at The Leaky Cauldron. Hogsmeade is the only all-wizarding village in Britain, and the home of Honeydukes sweetshop.

There are also places that exist within Muggle-populated areas, such as Sirius Black's house at Number Twelve, Grimmauld Place, set in a borough of London. Platform 9¾, where students board the Hogwarts Express, is within King's Cross Station, one of the busiest transportation hubs in the country—though these scenes were actually shot between Platforms 4 and 5.

Inspired by the many magical and amazing locations seen in the Harry Potter films, this signpost points the way to places that could be only a wand tap or an Apparition away.

Suggested locations include: Diagon Alley / Little Whinging / Godric's Hollow / The Burrow / Knockturn Alley / Hogwarts / Grimmauld Place / Shell Cottage / Hogsmeade / King's Cross Station / Budleigh Babberton / Forbidden Forest / Quidditch Pitch.

What you need

- 🥽
- 🧤
- 😷
- 🎧
- Pressure treated 4-by-4-inch piece of wood, 8 feet long
- Pressure treated 4-by-4-inch fence post cap
- Circular saw or handsaw
- Paintbrushes: a 1-inch flat brush; and angled and round paintbrushes
- White primer paint
- Paint
- Jigsaw
- 12 to 15 narrow boards, each about 1 to 3 inches tall and 1 to 2 feet wide. Old, used boards are better than new. Old fence pickets, lath, and wood scraps work best.
- Computer and desktop printer
- Cellophane tape
- Rags
- Drill and multiple-size bits
- Screw gun
- 1½-inch exterior wood screws
- Pencil
- Tape measure
- Floral wire
- 6-inch terra-cotta planter
- 2-inch bolt
- 2 large washers
- Shells, marbles, tiny terra-cotta pot, and other items to add to the boards
- Mailbox post holder spike

Instructions

1. Use the circular saw or handsaw to cut the 8-foot long 4-by-4-inch piece of wood down to the height you want for the space the sign will fill. The project photographed is 58 inches tall.
2. Use a paintbrush to paint the post with white primer paint.
3. Once dry, paint with a latex exterior house paint. It will most likely need two coats. The first coat should dry overnight before being painted over. This project uses a forest green.
4. Paint the pressure treated fence cap. These are available in the fence section of most hardware stores. Paint in a variety of colors if you have them.
5. Cut down the 12 to 15 boards to lengths of 1 to 2 feet wide.
6. Paint these boards in a variety of colors. Have fun! Add darker colors around the borders of some, or use contrasting colors on the edges and backs of the boards. Rub on wood stains, dry-brush paints, rub paint with rags, or splatter. Come up with your own techniques. There's no wrong here.

NOTE: This is a good project to use up paints and stains you have collected or stored from past projects. Exterior latex house paint is ideal, but any acrylic or house paints will work and give a vintage patina to the signs in a season.

7. Find stencils to use for destination names for this project at www.InsightEditions.com/HarryPotterHerbologyMagic.
8. Print destination names as large as you can on an 8½-by-11-inch sheet of paper. Cut out the letters.
9. Mark the middle of each line of type on your destination name printout. Note the middle of the board it will be painted on. Place the print on the board and tape down.
10. Paint the destination names. Use a soft bristle brush over the stencil. Choose a contrasting paint color to the background for each sign. The goal is to make the painted type very visible.
11. While you're painting destination name stencils, think about what items you have that might make nice add-ons for some signs—like adding shells for Shell Cottage, or a tiny broom for Hogwarts. Let your imagination wander!
12. Take the fence post cap, find the center of the top of the cap by drawing a pencil line from corner to corner. Drill a hole large enough for the 2-inch long bolt to go through. Add a washer to the bolt and push it through on the *underside* of the cap.
13. Use a small drill bit to pre-drill 2 screw holes on the top of the cap. Use wood screws and a screw gun or screwdriver to screw the cap onto the top of the pole. (You'll add the terra-cotta pot to the bolt that is sticking out the top as a last step, after the signs are attached to the pole.)

14. Lay the signpost on a table and lay out the signs on the pole in the order you want them. Pay attention to not have too many similar colors near each other, or to have too many signs of a similar length (or width) near one another. Consider angling some slightly to give the sign more dimension.

15. Once the order is set, use a small drill bit to pre-drill each board where you'll screw it into the pole. Pre-drilling will keep the screws from cracking and splitting the boards.

16. Use a screw gun or screwdriver to screw in each board, starting from the top. They don't have to be straight; some can be angled.

17. Once done, now is the time to add any decorative items to the boards. For this project shown, a tiny, plantable, terra-cotta pot was attached with a wire to the "The Burrow" sign; drilled holes on the "Ministry of Magic" sign held clear glass marbles, attached with hot glue on the back; a shell was added to the "Shell Cottage" sign; a tiny dollhouse broom was added to the "Hogwarts" sign; and a painted cutout circle with "9¾" was added to the King's Cross Station sign. Add your own items. Be creative!

18. Add the 6-inch terra-cotta pot to the top by putting it over the bolt sticking out from the top of the fence cap. With a large washer and nut, tighten the pot in place.

19. Find the perfect spot to "plant" your new Wizarding Wayfinding sign. Place a small piece of the 4-by-4-inch post into the mailbox post holder spike. Pound it into the ground as far as you're able so it can easily support your pole. Before you begin, make sure the post holder is facing the direction you wish. While pounding, make sure the post holder is upright and will hold your pole straight.

20. Place your pole into the holder.

21. Plant the terra-cotta pot on top and any attached to signs. Something spiky (like a grass) with something droopy (like a spider plant, silver falls, sweet potato vine, or autumn fuchsia).

Care

Ideally, the sign will last longer and colors at their best if it is placed in either shade or dappled shade. If you have severe winters where you live, storing it in a dry, covered location will add years to its life.

- **Maintenance:** As the sign ages, it will naturally get an aged patina. If after a few years, it starts to look like paints are fading quickly, a coat of clear polyurethane will keep it looking better longer.

DEATHLY HALLOWS TRELLIS

"The Elder Wand—the most powerful wand ever made. The Resurrection Stone. The Cloak of Invisibility. Together they make the Deathly Hallows. Together they make one Master of Death."

—XENOPHILIUS LOVEGOOD,
HARRY POTTER AND THE DEATHLY HALLOWS – PART 1

SKILL LEVEL: 4
TIME: 3 TO 4 HOURS

The Deathly Hallows are three legendary magical items that allow the possessor to become the "Master of Death." Voldemort, in his pursuit of immortality and power, seeks these out and even gets his hands on the Elder Wand. What he doesn't know is that Harry Potter has the Cloak of Invisibility and is later bequeathed the Resurrection Stone by Albus Dumbledore. When Harry defeats Voldemort in a wand battle, he acquires the Elder Wand—and promptly destroys it. He also drops the Resurrection Stone onto the floor of the Forbidden Forest. But there's always use for an Invisibility Cloak!

The Resurrection Stone was initially held in one of Voldemort's Horcruxes, the ring of his grandfather Marvolo Gaunt, which Dumbledore tried to obliterate. However, when the ring's stone was being created for *Harry Potter and the Half-Blood Prince*, designer Miraphora Mina did not know that it was etched with the sign of the Deathly Hallows, which was first revealed in the book *Harry Potter and the Deathly Hallows*. Fortunately, the seventh book came out before she finished it, and Mina could add the important image.

This trellis with the three signs of the Deathly Hallows seen in the films is a perfect place for climbing vines—it might even make you the master of greenery!

What you need

- 👓
- 🧤
- 🎧
- Tape measure
- Circular saw or handsaw
- 3 pressure-treated 2-by-2-inch-by-8-foot posts
- Four 5/16-by-48-inch round steel rods
- Three 14-inch floral hoops
- Drill and 3/16-inch and 3/32-inch bits
- Four 3-inch exterior wood screws
- Vise
- Hacksaw
- One 3/16-by-72-inch round steel rod
- Wire cutters
- Pliers
- Floral wire
- Epoxy
- Exterior latex house paint in color of your choice
- Two 2-foot-by-½-inch lengths of rebar
- Hammer

Instructions

1. Use the circular saw or handsaw to cut two 2-by-2-inch posts to 6 feet tall.
2. With the third post, cut a 25-inch piece. This will be your top piece.
3. Cut another board 22½ inches long. This will be the lower base.
4. Assemble the sides and top of the frame by attaching the top 25-inch piece to the two 6-foot-long sides. Pre-drill with a small drill bit, then screw one 3-inch wood screw into each board where they meet. The top board will cover the ends of the side boards. Don't put in the lower base yet. That will be last.

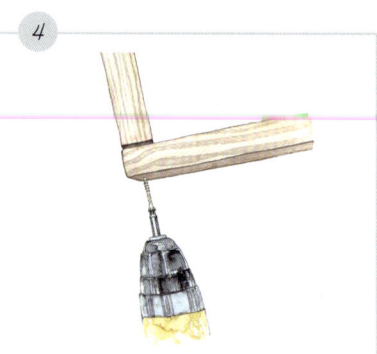

5. In the center of the top board, drill a hole with a 3/16-inch drill bit.
6. On the lower board, find the center, drill a 3/16-inch hole only *about halfway* through the board, not all the way through! Do not attach the lower board to the frame yet.
7. Take the 3/16-inch metal rod and feed it through the hole in the top board. Feed through to the partial hole in the lower board not attached to the frame.
8. Twenty-four inches down from the inside of the top board, on a side board drill a 5/16-inch hole.

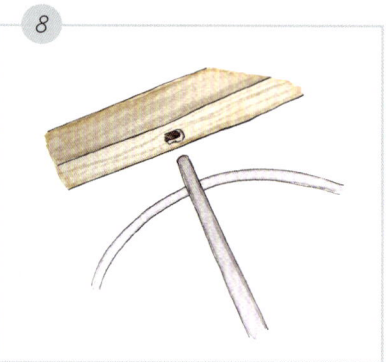

9. On the opposite side of the frame, drill a corresponding hole *only about halfway into the inside of the frame*.
10. Place one of the 5/16-inch metal rods in a vice and use a hacksaw to cut one of the rods in half to 2 feet each, carefully cutting near the vice itself to avoid bending the rod. Repeat with the other three 5/16-inch rods.
11. Lay the straight steel rod pieces out on a flat work surface with the floral hoops in the pattern of the Deathly Hallows mark.
12. Insert the rod into the hole in the frame on one side, fitting into the partial hole on the other side.
13. Starting with the top Deathly Hallows panel, drill a hole 22½ inches down from the first hole you drilled (48 inches down from the inside of the top board of the frame). Drill a partial hole on the opposite side, like above in step 8. Push in the second 2-foot section in the frame hole, resting in the partial frame opposite.

14. Just above where each rod fits into the sides of the frame, on both sides, you will drill a hole, at an angle that a $5/16$-by-24-inch rod will fit into. *Do not drill through the frame*, only partially. It should be angled so that the rod touches the center $3/16$-inch bar sticking through the top of the frame. The angle doesn't have to be perfect, but by placing the rod in position before drilling, you can see what angle it needs to be.

12 & 14

15. Do the same angled drilling to the two other Deathly Hallows panels below the top one.

16. Start assembly by connecting all metal rod and floral hoops with floral wire. Gloves are invaluable for working with floral wire. Wrap each point multiple times, crisscrossing and twisting with pliers for a tight connection and trim off excess.

17. For the joints where the tops of triangles meet the horizontal bars and the bottom of each circle, you'll have to do many wraps around, overlaying multiple times through and around until angled rods are secure. You will be creating a "cage" of floral wire to keep the rod ends together.

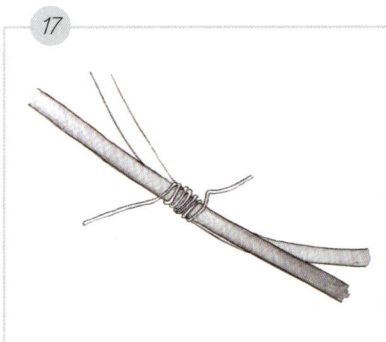

17

18. Now is time to add the bottom $22\frac{1}{2}$-inch board as the base to the lowest Deathly Hallows panel. Position at the bottom of the circle. It should be about 66 inches down from the top of the frame. No worries if it's not exact! Pre-drill and screw in the last two 3-inch exterior wood screws.

19. The $3/16$-inch metal rod will still be sticking out the top of the frame. With a hacksaw, cut that flush with the frame.

20. For the two horizontal bars, where they were inserted into the frame, add a dab of epoxy glue to secure them in place.

21. Paint the frame in a color of your choice. This frame was painted in a green paint used on other projects. Always shop your basement or garage (or those of your family and friends!) for leftover paints rather than buying, if you're able. Let dry overnight. You may need another coat.

22. On the bottom "feet" of the frame, drill, with a $\frac{1}{2}$-inch drill bit, holes in the center of each.

23. Place the 2-foot rebar pieces in each of the feet lightly. Put the frame in place in your garden, sticking the rebar off the bottom in the ground a few inches. Take the frame off the rebar.

24. Use the hammer to pound rebar stakes in the ground, then place the frame on the rebar. You may have to tap the frame at the top to get a tight fit on the rebar, and/or get the rebar deeper in the ground.

25. The Deathly Hallows Trellis is ready to be planted! See suggested plant list.

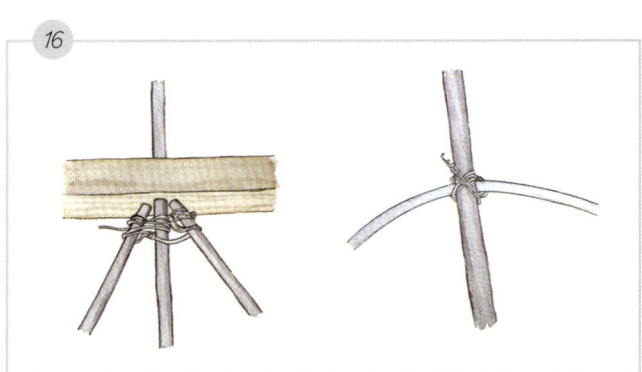

16

OUTDOOR DÉCOR

Plants

While it might be tempting to plant Devil's Snare on your Deathly Hallows Trellis, it is not recommended. Instead, try these non-threatening perennial vining plants:

- Clematis
- Akebia vine
- Passionflower
- Mandevilla
- Moonflower
- Honeysuckle

Care

The trellis should not need much maintenance at all. After a few years, the paint may start to wear, much like exterior house paint would. It may need repainting after 6 to 8 years.

Climbing Vines

Vines climb by different strategies.

- *Twining* vines climb by branches growing in a circling fashion, twisting themselves around and clinging to a support. These are vines like clematis, wisteria, black-eyed Susan vine, and Akebia vine.
(*Did you know Japanese wisteria twines clockwise and Chinese wisteria twines counterclockwise? Plants are amazing!*)

- *Tendril* vines have shoots that anchor themselves to a support as branches establish. Tendril vines examples are sweet peas, passionflower, grapes, and squash vines.

- *Suckering* pads attach to supporting structures. These damage surfaces and need to be carefully controlled. Sucker vines include English ivy and winter creeper.

GARDEN OF PROPHECY TRELLIS ORBS

"Prophecies can only be retrieved by those about whom they are made."

—LUCIUS MALFOY,
HARRY POTTER AND THE ORDER OF THE PHOENIX

SKILL LEVEL: 1
TIME: 1 HOUR

Harry Potter is drawn to the Ministry of Magic in *Harry Potter and the Order of the Phoenix*, when Voldemort projects a false vision into Harry's mind of his godfather, Sirius Black, being held there. Harry is joined by members of Dumbledore's Army, and find themselves in the Hall of Prophecy. Among myriad glass shelves packed with shadow-filled Prophecy globes, Neville spots one that has Harry's name on it—it is a prophecy by Sybill Trelawney concerning Harry and Voldemort.

The Hall of Prophecy was the first entirely digital set created for any Harry Potter film location, due to the intricacy and danger of (repeatedly) filming thousands of glass globes and shelves exploding and crashing to the floor. However, fifteen thousand orbs were manufactured and lit by the prop makers before the final decision to use CGI. Never one to waste a good prop, set designer Stephenie McMillan repurposed some of the globes as drink dispensers in the "Ministry Munchies" cart in the Ministry's Atrium.

The prophecy globes seen in the Ministry of Magic in *Harry Potter and the Order of the Phoenix* inspired this trellis, which will, predictably, be a fanciful way to show off a lush assortment of orb-supported vines.

What you need

- 🖐
- Four 10-inch floral hoops
- Floral wire
- Needle-nose pliers
- Wire cutters
- Four 19-inch floral hoops
- Four 14-inch floral hoops
- Regular pliers
- Sledgehammer
- 4-foot length of rebar

Instructions

1. Take 2 of the 10-inch floral hoops and set one perpendicular inside the other. Where they meet, wrap tightly with floral wire. Wrap the floral wire around and under a few times, with the wire crossing over itself a few times. Wearing gloves, twist tie the floral wire with needle-nose pliers when done. Use wire cutters to snip off any excess.

2. Add the next 10-inch hoop inside the 2 at an angle. Where it meets the 2 hoops, wrap with wire.

3. Add the fourth hoop at a different angle than the last one, creating few large voids in the sphere. Where it meets any other hoops, tie and twist on floral wire.

4. Repeat the same with the four 14-inch floral hoops.

5. Repeat the same with the four 19-inch floral hoops.

6. Take the 4-foot length of rebar, and pound it into the ground with the sledgehammer, leaving about 3 feet above ground.

7. Put the spheres over the rebar stake.

8. Plant something viney at its base. See suggested plant list.

Plants

While it might be tempting to plant some Devil's Snare on your Prophecy Orbs, it is not recommended. Instead, try these non-threatening perennial vining plants:

- Clematis
- Passionflower
- Mandevilla
- Moonflower
- Honeysuckle

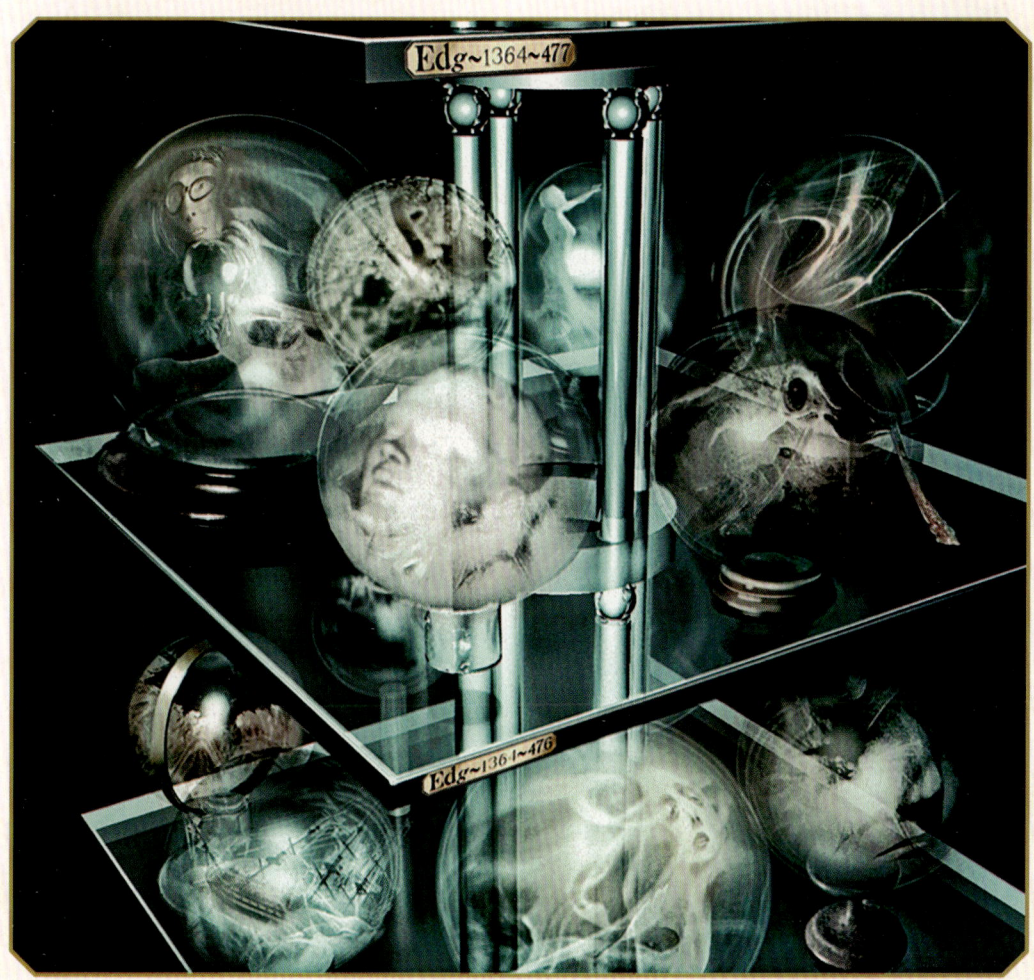

SORTING HAT HANGING BASKET

"When I call your name, you will come forth. I shall place the Sorting Hat on your head, and you will be sorted into your houses."

—PROFESSOR MINERVA MCGONAGALL,
HARRY POTTER AND THE SORCERER'S STONE

SKILL LEVEL: 3
TIME: 3 TO 4 HOURS

At the Sorting Ceremony for first years at Hogwarts, a conical hat is placed atop each new student's head to announce into which house they will be placed. Some students are easily placed—Ron is the sixth Weasley child to be placed in Gryffindor, and Draco Malfoy follows in his father and mother's footsteps. Harry Potter was a harder decision for the hat, who thought he could be placed in Slytherin. But, as Harry tells his son, Albus, many years later, the hat will take your own choice into consideration, and Harry is sorted into Gryffindor.

The Sorting Hat seen through the Harry Potter films was created in both digital and practical constructions. For practical uses, first a leather material was squished down into a cone shape, soaked in hot water, and then dried on a heating unit. Wires were inserted to keep its shape, and some were lined with a horsehair canvas for stability. The hats were dyed, aged, and imprinted with Celtic symbols. Seven hats were created, and though they may have had different wrinkles or shading, you never see them together!

Hang this Sorting Hat–shaped planter outside your own house, whether you're a Gryffindor, Hufflepuff, Ravenclaw, or Slytherin.

What you need

- Cone-shaped hanging basket, with coco fiber liner
- Corrugated plastic panel at least 24-by-24 inches, like a used realtor lawn sign (reuse/repurpose/recycle!)
- Tape measure
- Flat stick about 12 inches long
- Drill and an 1/8-inch and a 7/32-inch bit
- Black marker
- One nail
- Hammer
- Flat board
- Utility knife
- Scissors
- Heavy-duty sewing thread
- Landscape fabric
- Sewing machine
- Floral wire
- Plants (see Plants below)

NOTE: This is an exterior hanging basket. Excess water will seep through and drip to the ground.

Instructions

1. Find a wire formed cone-shaped hanging basket (available at most large garden centers or online). The one used in this project is 12 inches wide and 19 inches tall. They usually come with a coco moss insert with an apparatus for hanging. Remove the coco moss insert and hanging wires or chains (you'll add them both back in later).

2. Make a brim for the hanging basket with the corrugated panel first by finding the center of the 24-by-24-inch square panel. This will be the center mark where you'll make a compass to make a perfect circle.

3. Drill 2 holes 1 inch from each end of the 12-inch stick. One hole should fit the nail (1/8-inch bit); the other should fit the black marker (7/32-inch bit).

4. Place the corrugated panel on a work surface or a large flat board that you'll be able to lightly hammer a nail in temporarily for the compass.

5. Place the 12-inch stick onto the corrugated panel and hammer the nail through the nail hole on the stick into the center mark on the corrugated panel. Pound in lightly; you'll be pulling it out shortly.

6. With a marker in the marker hole 10 inches away from the center of the corrugated panel, rotate to make a circle.

7. With a marker in the hole 5 inches from the center of the corrugated panel, rotate to make another circle.

8. On a work surface, with the utility knife, cut the corrugated panel's inner and outer circles.

9. This is now your brim form. Cut a line through the inner circle to the outer circle. This will make it easier to fit inside the sewn pieces later on.

10. Lay the circle shape on the landscape fabric. Use the scissors to cut the landscape fabric, allowing for up to ½-inch extra fabric on both the inside and outside edges. Cut out the inner circle leaving at most ½ inch of fabric for sewing. Make two of these. You'll end up with two large donut-shaped pieces of landscape fabric.

11. Using leftover scraps from the center cutout piece of fabric, cut two 1½-inch-wide strips of fabric.

12. Leave one of the donut shaped fabric pieces intact. On the other, cut an opening from the outside to the center. Place a 1½-inch fabric strip on either side of the cut, and sew the strips to the main pieces of fabric.

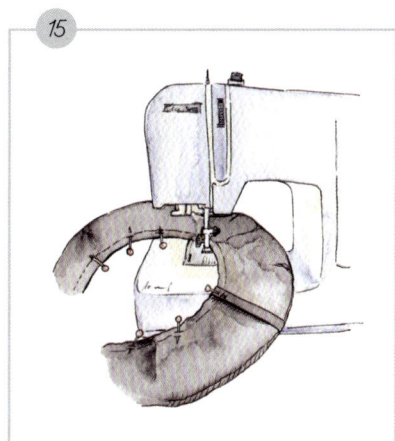

13. Using a finger, vigorously rub along the seam line, pushing the two seams open. Do not use an iron, it will melt the landscape fabric.

14. Lay the two pieces of fabric on top of each other, with the two added pieces on top and on the outside. Pin around the outside edge.

15. Lay the brim form on top of the two pinned pieces. With a marker, draw a circle 1 inch from the outer edge. Using a sewing machine, sew around the circle. The two added pieces will overlap a bit.

16. Remove the pins and turn the piece inside out, and hand press around the outside edge.

17. Slide the corrugated brim form into the opening in the fabric. Continue slowly sliding the brim into the opening until it is completely enclosed.

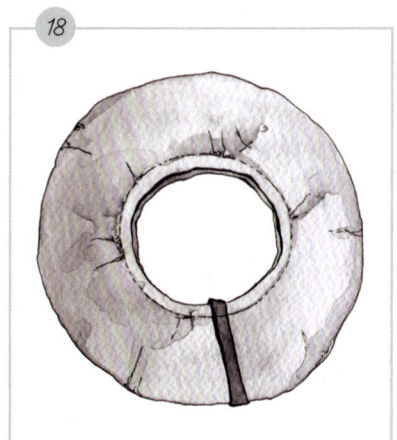

18. Pin the two pieces of the inner circle together and sew around the edge of the inner circle as close to the brim form as possible. This will give the fabric a snug fit. Remove the pins and neatly trim with scissors.

OUTDOOR DÉCOR

19. Cut a piece of landscape fabric to fit the cone. Lay the cone on the landscape fabric and roll to cover the cone. Loosely cut the shape that fits the cone. Make sure it's at least 2 inches longer than the cone and 2 inches wider than the cone.
20. Lay the fabric down and fold the diagonal edges to meet. It will look a bit like a dunce cap.
21. Pin along the diagonal edge and sew the length of the edge. Remove pins and turn the piece inside out.

19–21

22. Place the fabric cone over your cone. Fold the top edge around the opening at the wide end of the cone toward the inside of the cone. Pin all the way around the edge.
23. Using a needle and thread, sew as close to the top of the cone to hold the fabric in place. An in-and-out stitch is sufficient, every ½ inch, all the way around.

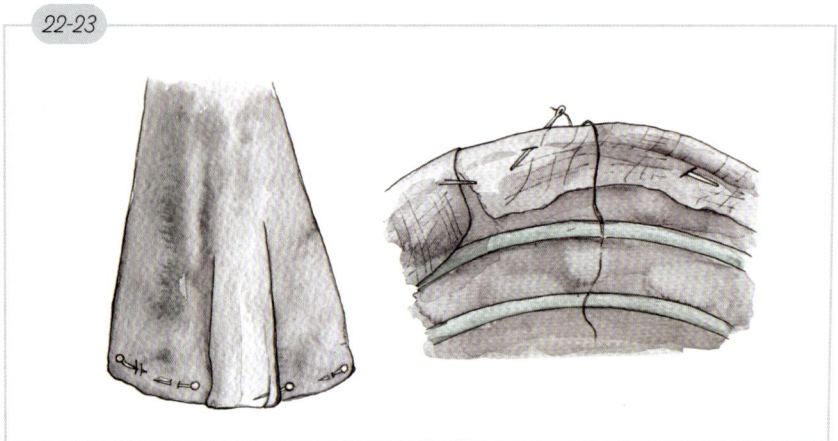
22–23

24. Remove the pins. With scissors, trim the extra fabric to 1 inch around the inside of the cone.
25. Attach the covered brim form to the covered cone shape with floral wire in at least four spots by inserting through the landscape fabric and corrugated panel and twist-tying together to the wire of the cone basket.
26. Add the coco fiber liner and attach the hanging chain or wire.
27. Plant up the hat with a thriller, a filler, and a spiller!

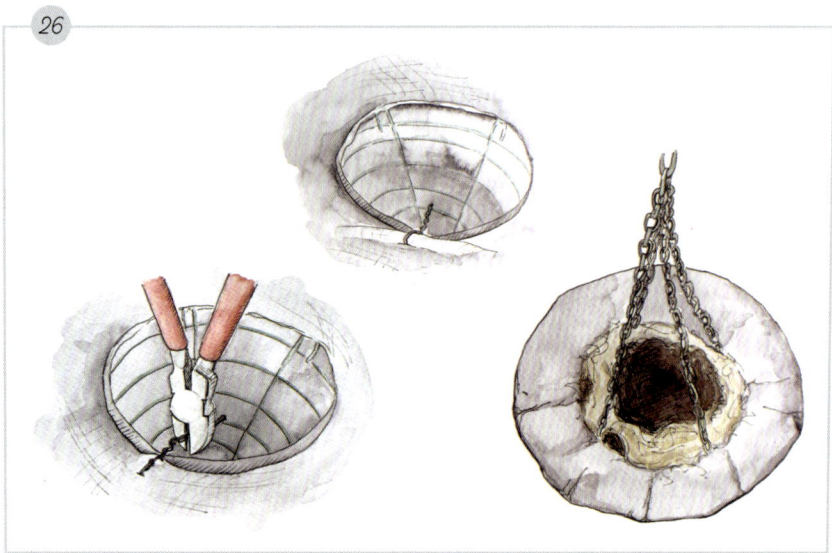
26

Plants

Any plant store, garden center, or big box hardware store will have suitable plants for outdoor use. Ask a plant store professional for recommendations! Some suggested plants:

Thrillers:

- Purple fountain grass (*Pennisetum setaceum* 'Rubrum')
- Ferns
- Persian shield (*Strobilanthes dyerianus*)
- Blue storm lily of the Nile (*Agapanthus praecox orientalis* 'Atiblu')
- Ornamental millet
- Hellebores (*Helleborus orientalis*)

Fillers:

- Petunia
- Polka-dot plant (*Hypoestes phyllostachya*)
- Alyssum
- Coleus
- Asparagus fern (*Asparagus densiflorus*)
- Begonias
- Impatiens

Spillers:

- Variegated ivy
- Sweet potato vine (*Ipomoea batatas*)
- Creeping Jenny (*Lysimachia nummularia*)
- Trailing lantana (*Lantana montevidensis*)
- Silver falls (*Dichondra argentea* 'Silver Falls')
- Variegated creeping Charlie (*Glechoma hederacea*)
- Autumn fuchsia (*Autumnale Fuchsia*)
- Vinca vine (*Vinca major*)
- Creeping licorice (*Helichrysum petiolare* 'Limelight')

Care

These are normal annual plants requiring regular attention.

- **Light:** Direct sun is best for flowering plants and grasses. Minimally they need bright sunshine, all day, outdoors.
- **Water:** Near daily watering may be needed for hanging baskets of plants in summertime. Droopy leaves mean they need more water.
- **Maintenance:** Always trim dead leaves to keep the plants looking their best.

FORBIDDEN FOREST FENCE PLANTER

"The forest? I thought that was a joke. We can't go in there. Students aren't allowed. And there are . . . werewolves!"

—DRACO MALFOY,
HARRY POTTER AND THE SORCERER'S STONE

SKILL LEVEL: 3
TIME: 2 TO 3 HOURS

Create your own "forbidden" forest with this unique planter box that attaches to any fence and features labels you can use to call out the plants seen in the Harry Potter films. The Forbidden Forest (also known as the Dark Forest) was seen in every film except for *Harry Potter and the Deathly Hallows – Part 1*.

For each appearance, production designer Stuart Craig challenged himself to make the forest bigger, darker, and more mysterious, with thicker mists and more intimidating flora. For *Harry Potter and the Order of the Phoenix*, Craig redesigned the trees' roots after tropical mangrove trees, which looked to him as though the trunks were supported by fingers. This version of the trees was bigger than the redwoods in Northern California, with circumferences of twelve to fourteen feet around. The biggest forest was built for *Harry Potter and the Deathly Hallows – Part 2*, with a six-hundred-foot-long cyclorama backdrop that circled the set. Your forbidden forest need not be as big, but it will definitely add some magic to your garden.

What you need

- 😷
- 🎧
- Tape measure
- Circular saw or handsaw
- Black marker
- One 2-by-8-inch-by-8-foot pressure treated board
- Drill
- 1-by-4-inch-long flat stick
- Hammer
- One nail
- Drill bit, larger than the jigsaw blade is wide
- Jigsaw
- Sandpaper
- Paintbrush
- Chalk paint (optional: exterior house paint or stain)
- 6½-inch exterior wood screws

Instructions

1. Use the handsaw to cut the pressure-treated board down to 4 feet. Most home stores that sell lumber will cut it for you for free (so you may not even need that circular or handsaw!). With one 8-foot board, you can make two of these racked plant hangers!

2. Mark the center lengthwise and widthwise. That's the center of your first planter hole.

3. Draw a line lengthwise on the centerline.

4. Measure out 9½ inches from the center of the board along the line drawn along the center. Use the black marker to mark that spot with an X.

5. Measure another 9½ inches from the X and mark that spot with another X. Those are your planter hole centers for one side. Now do the same on the other side of that first centered X, with X marks 9½ inches apart.

6. To make a compass, drill holes 2¾ inches apart on the 4-inch stick. One hole should fit the nail, the other should fit the black marker.

7. Use the hammer and nail to nail the stick lightly to one of the center marks on the board.

8. Put the marker in the marker hole and drag the 4-inch stick to make a circle. This is your cutting mark. Remove the marker, nail, and stick.

9. Repeat with nailing and drawing a circle on the other four planter holes.

2-5

7-9

10. With a large drill bit, larger than the jigsaw blade is wide, drill a hole along the inside of each marked circle.

11. With a jigsaw, cut out the circle of each of the five planter holes.

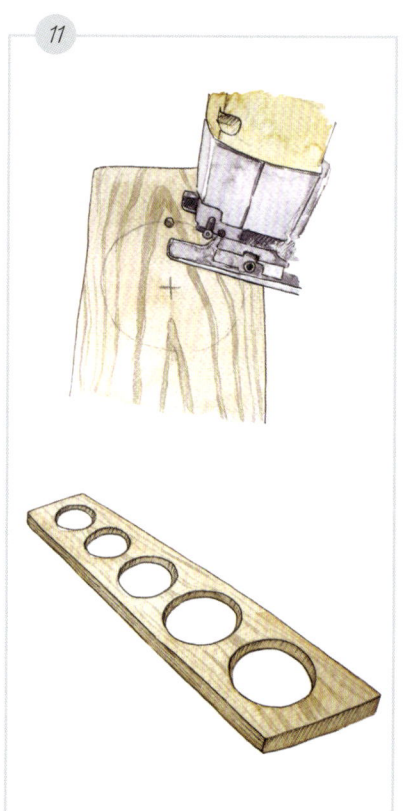

12. See how the pots fit in the holes. Some extra cutting may be necessary or smoothing out spots with sandpaper.

13. Sand the ends of the board.

14. Paint with chalkboard paint if you want to be able to write plant names. Otherwise, you can use exterior house paint or wood stain of your choosing.

15. This plant holder is intended to fit on a standard picket fence along a mid-height horizontal support. Using the 6½-inch exterior wood screws, screw along the board to the fence's horizontal support.

16. In chalk, write out names for the plants on the front of the board, but give them magical names (not Muggle names!), like Devil's Snare, Wolfsbane, Puffapod, Scurvy Grass, Dittany, Flutterby bush, Bubotuber, Alihotsy, Flitterbloom, or Mimbulus Mimbletonia. Magical gardens are always more interesting than Muggle gardens.

Plants

Any pot-growing plants will find a happy home in this Forbidden Forest plant holder. Some suggestions:

- Geraniums
- Petunias
- Mandevillas
- Chrysanthemums
- Begonias
- Coral bells (*Heuchera sanguinea*)
- Impatiens
- Hosta
- Creeping Blue Blossom (Creeping Ceanothus) *Ceanothus thyrsiflorus*

Care

These are plants requiring regular attention.

- *Light:* Flowering plants will like direct sunlight. Bright light will be necessary for any outdoor plant. Read plant labels carefully. You'll have more success (and less work) if you use plants with similar needs planted together.

- *Water:* Plants in pots will dry out more quickly than plants in the ground. Monitor plants consistently. They'll need watering possibly every other day—less when it rains and more when there's a dry spell.

- *Maintenance:* Always trim dead leaves to keep the plants looking their best.

OUTDOOR DÉCOR   133

DIRIGIBLE PLUMS SIGN

"Xenophilius Lovegood. We live just over the hill."

"Pleasure to meet you, sir."

—XENOPHILIUS LOVEGOOD TO HARRY POTTER,
HARRY POTTER AND THE DEATHLY HALLOWS – PART 1

SKILL LEVEL: 2
TIME: 2 TO 3 HOURS

In addition to the Butterbeer necklace Luna Lovegood wears in *Harry Potter and the Order of the Phoenix*, she also sports a pair of earrings that resemble Dirigible Plums—an orange-colored fruit that grows upside down. A bush of these gravity-defying plums, which more actually resemble radishes, surround the Lovegood house, fronted by a large sign that cautions visitors to keep off the Dirigible Plums.

As arts-and-crafty as her character, actress Evanna Lynch helped design Luna's beaded earrings, honing the shape and color with costume designer Jany Temime. Lynch also made a beaded hare bracelet and contributed to the design of the lion hat she wore as a supporter of the Gryffindor Quidditch team in *Harry Potter and the Half-Blood Prince*.

It should be no surprise that Luna would be influenced by her home environment in creating the Dirigible Plum earrings. With Dirigible Plums made from ping-pong balls "floating" in place, this sign based on the film offers a fun warning to your visitors as well!

What you need

- 🥽
- 🧤
- 😷
- 🎧
- Pencil
- Tape measure
- Handsaw or circular saw
- 1 cedar fence picket, ⅝-by-6-inch-by-6-foot
- One 2-by-2-inch-by-8-foot post, pressure treated
- Drill and various small bits
- Screw gun
- Two 2-inch exterior wood screws
- Hammer
- 8 galvanized exterior nails
- Sandpaper
- Paintbrushes
- Computer and tabletop printer
- Cellophane tape
- Dark orange exterior house paint
- Vice
- Hacksaw
- One ⅛-inch-by-36-inch round rod
- Two ping-pong balls
- Floral tape
- Sponge
- Orange paint
- White paint
- Glue gun
- Fake leaves

Instructions

1. From the cedar fence picket, use the handsaw or circular saw to cut two 2-foot panels.
2. From the pressure treated 2-by-2-inch post, use the handsaw or circular saw to cut 2 pieces at 19 inches (the legs), and cut a third at 23 inches long (the top).
3. Mark off 2 inches from each end and centered on the 23-inch post. Attach the two 19-inch posts perpendicular to the 23-inch piece at the 2-inch mark: Pre-drill one hole on each mark on the ends of the 23-inch post with a drill bit just smaller than the wood screws. Use the screw gun or a screwdriver to screw the 2-inch wood screws into the posts where you pre-drilled.
4. With the U-shaped post structure flat on the table, place the first cedar fence picket parallel on the face so that the edge of the cedar picket is flush with the top of the 23-inch post and centered on the U-shape (you should have a ½-inch overhang on each end).
5. With the galvanized nails, hammer nails to the top cedar fence picket on the 19-inch posts (legs) in two spots an inch from the top and bottom of the cedar picket.
6. Attach the second cedar picket just below the first one in the same way.

TIP: Pre-drilling the nail holes with a bit slightly smaller than the nails will minimize the risk that the cedar fence panel will not crack or split.

7. Use the sandpaper to sand all edges.

8. Find a stencil to use for this project at www.InsightEditions.com/HarryPotterHerbologyMagic.

9. Print out the stencil on a desktop printer. It will take a few pages if you can only print 8½-by-11-inch paper. Cut out the letterforms. Note the center of the printed out text. Mark the center of the cedar panels.

10. Find the center of the stencil and the center of the sign panel.

11. Paint the text with a soft-bristle brush in the dark orange paint. Cedar planks are rough, so you may have to paint two coats for good coverage.

12. With an ⅛-inch drill bit, drill 2 holes 1½ inches deep on the top center of the 23-inch post, 3½ inches and 6 inches in from the right side. Make sure the holes are as straight as possible.

13. Place the ⅛-inch metal rod into a vice and saw the rod to get two pieces—one 18 inches long, the other 12 inches long. TIP: Placing your cut point as close to the vice as possible will help reduce bend in the metal bar as it's being cut.

14. Place the metal rods into the drilled holes along the top of the sign. A light tap with a hammer should make them snug in the holes. Make sure they are straight and true.

15. With floral tape, wrap the metal rods.

16. Drill a hole with a small drill bit on one end of each ping-pong ball but do not drill all the way through. This is now the bottom of the ping-pong ball.

17. Paint the ping-pong balls with the white and orange paints, blending the paints from orange on top to white on the bottom to resemble a Dirigible Plum. A sponge will help you blend colors. Let dry.

18. Once dry, place on the metal rods, inserting the metal rod into the hole in the ping-pong ball.

19. Add fake leaves to the bottom of each rod with the hot glue gun. Cover hot-glued area with floral tape.

20. Drill a small hole in the very top of the ping-pong ball, add a drop of hot glue, and insert fake leaves into the newly drilled holes.

Care

Ideally, the sign will last longer and colors at their best, if it is placed in either shade or dappled shade. In cold climates, it should be brought inside in winter.

- *Maintenance:* As the sign ages, it will naturally fade to gray. If after a few years, it starts to look like paint is fading, it may need to have paint refreshed.

OUTDOOR DÉCOR 139

PRIVET DRIVE HOUSE NUMBERS

"Albus, do you really think it's safe? Leaving him with these people. I've watched them all day. They're the worst sort of Muggles imaginable."

—PROFESSOR MINERVA MCGONAGALL,
HARRY POTTER AND THE SORCERER'S STONE

SKILL LEVEL: 1
TIME: 1 HOUR

For the first eleven years of his life and then each summer between years at Hogwarts, Harry Potter lived with his only living relatives—his Uncle Vernon, Aunt Petunia, and cousin Dudley, at Number Four, Privet Drive, a house the Dursleys are incredibly proud to *not* have stand out in their neighborhood.

The house, first seen in *Harry Potter and the Sorcerer's Stone*, sits in what is called a "ribbon development," house after house after house looking the same, which is what the Dursleys desire—to fit in with everyone else. Director Chris Columbus wanted a place that would appear oppressive, a "mean, awful place," as he described it. But anyone who works with plants or tends gardens indoors or out would want the complete opposite from the films—desiring their house to stand out, and this ingenious planter calls out the house owner's individuality and pride.

What you need

- 15-by-7-by-2-inch shadow box frame
- Glue gun
- Preserved moss
- Screwdriver
- House numbers, 4 inches tall

Instructions

1. Open box and separate back panel.
2. With hot glue gun, glue preserved moss onto the back panel of the shadow box.
3. Use the screwdriver to screw in house numbers on top of the moss and into the back panel of the frame. Be careful not to screw too far through the back panel. Close up frame.

Care

Keep out of direct sun and sheltered from severe weather.

PENSIEVE BIRD AND BEE BATH

"It's a Pensieve, very useful if, like me, you find your mind a wee bit stretched."

—ALBUS DUMBLEDORE,
HARRY POTTER AND THE GOBLET OF FIRE

SKILL LEVEL: 1
TIME: 45 MINUTES

In *Harry Potter and the Goblet of Fire*, Harry discovers that Albus Dumbledore keeps a Pensieve tucked away in a cabinet in his office. This magical device allows the user to view memories—theirs or others—"extracted" typically by a wand touched to their head, demonstrated by Albus Dumbledore and Horace Slughorn. But these memories can be tampered with, as Slughorn did to conceal his memory of Voldemort asking for information on Horcruxes. These remembrances can even come from tears, seen when Severus Snape wept out his memories for Harry to view in the Pensieve, where he learns the truth about Snape and his own destiny.

For the first appearance of the Pensieve, digital artists created a complex water surface with threads of silver fluid floating in the shallow basin that morphed into the memory when Harry was pulled into it. For *Harry Potter and the Half-Blood Prince*, the Pensieve bowl was suspended in midair. Once the memories were poured in, inky black threads were followed downward by the viewer and formed into the elements of the memory.

This Pensieve-style birdbath, inspired by the one seen in Dumbledore's office in the films, is for those with too many thoughts about their gardens. (Bees will like it, too.)

What you need

- 16-inch terra-cotta planter saucer
- 1½-inch paintbrush
- Silver acrylic paint
- Wire cutters
- Tomato cage (optional)
- Construction adhesive (optional)
- Clear marbles

Instructions

1. Make sure the terra-cotta planter saucer base is clean.
2. Use the paintbrush and acrylic paint to paint the interior of base silver. Allow to dry completely.
3. Once dry, find a good spot in your yard or garden. Near a tree but not under one is ideal. Birds and bees will want a safe haven tree to fly off to if they feel threatened. Find a level, dappled-sun spot where birds can easily see it and where it won't collect leaves, dirt, or debris. Debris can cause mold and bacteria which can harm birds. Make sure it is not likely to tip, spill, fall, or lean.
4. With the wire cutters, cut down a tomato cage so the Pensieve can sit on top and the "legs" can be inserted into the ground until level and stable. The Pensieve can alternatively be hung from a tree, set on a pedestal, stack of bricks, stump, a large terra-cotta pot set upside down, or an old stepladder. Whatever you use for a base, make sure it is secure and not wobbly. A dab of construction adhesive may be necessary.
5. Fill with clean water regularly—rainwater is best.
6. Add the marbles to make it easier for bees and other pollinating insects to enjoy a drink.

OUTDOOR DÉCOR 147

GREAT HALL FLOATING CANDLE LIGHTS

"It's not real, the ceiling. It's just bewitched to look like the night sky. I read about it in Hogwarts: A History."

—HERMIONE GRANGER,
HARRY POTTER AND THE SORCERER'S STONE

SKILL LEVEL: 3
TIME: 2 HOURS

Inspired by Hogwarts Great Hall as first seen in *Harry Potter and the Sorcerer's Stone*, you can add a Hogwarts Great Hall ambiance to your garden with these floating candle lights. In the Great Hall, a magical ceiling provides views of the day and night skies, and beneath that float hundreds of candles to illuminate the room.

For the very first time shooting inside the Great Hall, the filmmakers suspended 370 real candles, all floating up and down at different times. These candles hung from wires, but when it was learned the wires would break after only an hour due to drafts blowing the flames onto them, digital versions were created that were not only safer, but could also allow for imaginative arrangements, such as spirals or arches.

Dame Maggie Smith, who plays Professor McGonagall, remembers walking into the Great Hall the first time: "It was just so amazing. And it still has that effect on me, when the candles are moving up and down; it's a magic place to be." Your friends and family will have the same reaction to this enchanting illumination.

What you need

- 🥽
- 😷
- 25-light, 24-foot strand of white LED lights 1 foot apart
- Tape measure
- Pencil
- Two 10-foot-by-½-inch PVC pipes, white
- Vice
- Hacksaw
- Handheld mouse sander

Instructions

1. Using a vice and the hacksaw, cut the two 10-foot PVC pipes into 9-inch lengths, to yield 26 pieces. It cuts very easily, better with a hacksaw than with a power tool.
2. On one piece, mark 1 inch down from the top. With the hacksaw cut only *halfway* into the pipe.
3. Mark ½ inch below that cut.
4. From that ½-inch mark, mark at a 45-degree angle, cut up toward the last cut, ending where that first cut ends, halfway into the pipe. You should have a wedge-shaped notch in the pipe.
5. From the top of the pipe, cut one line from the top to the notch. When the candle is upright, this will be a vertical cut from the top to the notch.
6. With the mouse sander, sand off rough edges and any writing on the pipe.
7. Repeat this until you have twenty-five 9-inch pieces made.
8. Each LED light will have four strands. Take the two strands that connect to the LED bulb and squeeze them in the vertical cut, between top and notch. You may have to stretch the vertical cut a bit to fit them in. The two strands not connected to the LED bulb will nestle inside the notched cutout.
9. Repeat this on all the lights until you have them complete.
10. Hang outdoors (or indoors!). Note that these lights may have to be hung higher than normal because of the size of the "candles."

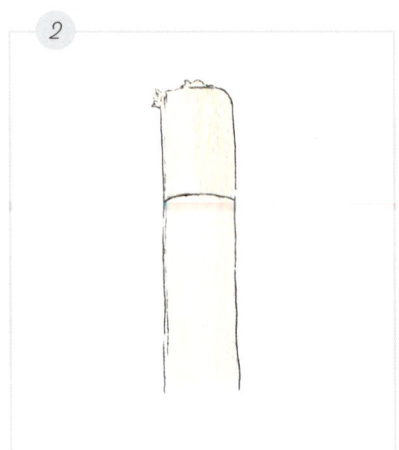

2

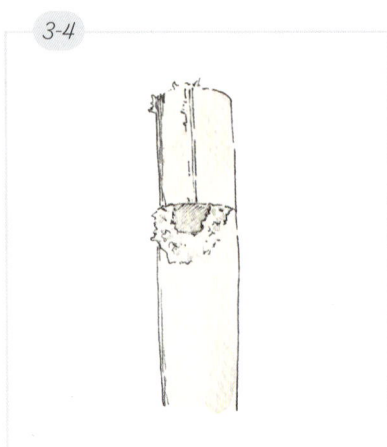

3-4

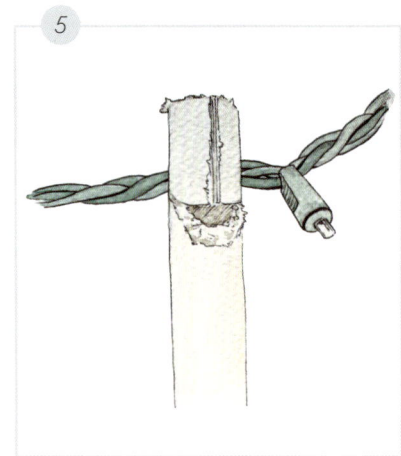

5

8

OUTDOOR DÉCOR 151

TREVOR THE TOAD'S ABODE

"Has anyone seen a toad? A boy named Neville's lost one."

—HERMIONE GRANGER,
HARRY POTTER AND THE SORCERER'S STONE

SKILL LEVEL: 1
TIME: 1 HOUR

First-year students to Hogwarts may bring, if they desire, a companion animal such as an owl, a cat, or a toad. Harry Potter has Hedwig, his owl; Hermione Granger eventually brings her cat, Crookshanks; and Neville Longbottom has his toad, Trevor, first seen in *Harry Potter and the Sorcerer's Stone*, or has him most of the time—the Gryffindor student seems to have a habit of losing his toad at inconvenient times. It might have helped Neville to have a toad abode for Trevor, which inspired the idea for a residence such as this one.

Four toads essayed the part of Trevor for the films; the real name of the one most frequently seen onscreen was, coincidentally, Harry! When not on a set, the toads lived in large, moss-based aquariums that were specially heated for their comfort. For Trevor's scenes, a trainer would either hand the toad to actor Matthew Lewis or place it on a prop (like an armchair or stair step) and retrieve it immediately afterward when the shot was complete.

Any garden-variety toad or frog will find this abode a safe and comfortable place—there will be little roaming from this home!

What you need

- 🥽
- 6-inch terra-cotta pot
- Handheld jigsaw
- Acrylic paint
- Paintbrush, thin, angled, or round
- 6-inch terra-cotta pot saucer
- Superglue
- 10-inch terra-cotta pot saucer
- Pebbles and stones

Instructions

1. Have the terra-cotta pot upside down. Wearing safety glasses, use the jigsaw to slowly cut a doorway notch (4-by-3-inch) out of the side of the terra-cotta pot, along the lip of the pot. Go slow, as not to let the terra-cotta pot crack.
2. Above the doorway, use paintbrushes to paint Trevor's name, (you may come up with clever names on your own). Painting "Welcome" is friendly, but remember, not all toads can read.
3. Glue the 6-inch terra-cotta pot saucer right-side up to the top of the upside down pot.
4. Find a good spot for the toad house. They need to be protected from the sun and from predators. Near a flower or vegetable garden will be good—take advantage of their insect-eating appetites! Toads will like quiet, shady spots away from your garden's foot traffic. Under trees with other low growing plants is ideal. Cats and dogs should be kept away if at all possible.
5. If possible, position the pot so the entrance faces north, away from the sun.
6. Add some dirt and leaves to the floor of the toad house.
7. About 6 inches in front of the toad house, dig a shallow hole in the ground to place the 10-inch terra-cotta pot saucer. Dig it shallow enough that the saucer's rim is ground level. Toads don't drink, they absorb water through their skin. They'll need the saucer, filled consistently with water, to want to stay near their new home.
8. Place pebbles and stones around the saucer to help it blend in.
9. Place a flat stone or pebble in the saucer and close to the edge to help a toad climb out of the saucer easily. They'll have no problem getting in, but they may have trouble getting out without different levels to climb.
10. In the glued 6-inch plant saucer, add a few plants found around where you placed the toad house. Native plants that grow naturally are best.

Toad Tidbits

- Toads are great additions to any garden. They will eat thousands of insects each month—flies, mosquitos, grasshoppers, ants, crickets, beetles, earwigs, wasps, and even snails and slugs!

- A toad's skin is very porous. It's how they hydrate. They are very sensitive to any type of pesticides, even organic ones. They can be harmful to toads.

- They usually live in holes in the ground or under logs or rocks. They can live farther from water sources than frogs can.

- They live in one area their entire life. Wild toads can live 10 to 12 years. In captivity they can live 30 years or longer!

- They are most active at night.

- Toads are completely harmless—they do not cause warts!

- Toads are native to all parts of the world except Australia, Greenland, Madagascar, New Guinea, New Zealand, and polar regions.

Chapter 4

EVERY GARDEN IS MAGIC

······◆······

""You know, if you're interested in plants, you should use <u>Goshawk's Guide to Herbology</u>. There's someone in Tibet who's growing gravity-resistant trees . . ."

—NEVILLE LONGBOTTOM, *HARRY POTTER AND THE GOBLET OF FIRE*

WIZARDING WORLD PLANT SUBSTITUTES

"Now, as our Mandrakes are still only seedlings, their cries won't kill you yet, but they could knock you out for several hours."

—PROFESSOR POMONA SPROUT,
HARRY POTTER AND THE CHAMBER OF SECRETS

Plants of the Wizarding World can be helpful or harmful, and sometimes even both. Mandrakes—that screech and squirm in their infantile stage—are used to create the Mandrake Restorative Draught, used by Professor Sprout to bring back those who have been Petrified by the Basilisk in *Harry Potter and the Chamber of Secrets*.

- The sound designers did, indeed, use the cry of a hungry baby as a start to creating the Mandrake's ear-splitting scream.

- In *Harry Potter and the Goblet of Fire*, Neville Longbottom offers Harry Gillyweed to succeed in the second task of the Triwizard Tournament.

- The "Gillyweed" Daniel Radcliffe ingested was made from black licorice—one of the actor's least favorite foods.

There are many strange and interesting plants in the magical world—as there are Muggle plants with shapes and textures that actually bear a resemblance or bring to mind those in the world of Harry Potter. When selecting plants you think might be a great substitute for a wizarding plant, just make sure they won't be harmful to you or your pets. Following are some magical plants and their non-magical Muggle equivalents for these projects or your garden.

✧ IN THE WIZARDING WORLD:
Gillyweed

Seen in *Harry Potter and the Goblet of Fire.* A native of the Mediterranean Sea, Gillyweed is best known for its gill-growing abilities on witches and wizards. But it has to be eaten raw, and it looks like slimy gray-green rat tails. So, good luck!

✧ IN THE MUGGLE WORLD:
Dancing-Bones Cactus
HATIORA SALICORNIOIDES

This Gillyweed substitute is the opposite of Gillyweed—it's a shrubby cactus native to Brazil that doesn't like a soggy bottom. In its natural setting, it's an epiphyte, which means it doesn't grow in the ground. It grows on tree trunks. It goes dormant in the winter months and wants even less water then

✧ IN THE WIZARDING WORLD:
Mandrake

Seen in *Harry Potter and the Chamber of Secrets.* Mandrake is used in restorative potions and has a legendary ear-piercing squeal that can knock you out for hours. It can be dangerous if not handled properly—or in the right hands.

✧ IN THE MUGGLE WORLD:
Ficus Bonsai
FICUS RETUSA

Ficus Bonsai, which is not as dangerous as a Mandrake, will give you the look and feel of a Mandrake, without the noise. It's an "indoor" tree that does not like frosty conditions. It's great as a bonsai because of its small size and thick trunk.

❖ IN THE WIZARDING WORLD:
Venomous Tentacula

Seen in *Harry Potter and the Half-Blood Prince.* Green, spiky, fanged, venomous, deadly, and with a poisonous juice, the Tentacula is often kept as a pet. Its leaves are a high-valued potion ingredient, fetching good prices according to leaf-pilfering Horace Slughorn.

❖ IN THE MUGGLE WORLD:
Ruby Glow Peperomia
PEPEROMIA GRAVEOLENS

With softer edges and no venom, this succulent, native to the cliffs of Ecuador, makes a much better pet. Its wine-red stems and leaves can produce rat-tail looking flowers. Be warned though, the "graveolens" in its Latin name means "bad-smelling."

❖ IN THE WIZARDING WORLD:
Whomping Willow

Seen in *Harry Potter and the Chamber of Secrets* and *Harry Potter and the Prisoner of Azkaban.* Probably Wizarding World's most notorious horticultural star, this deciduous bully will attack anything within reach of its canopy. This is why it's great for covering up secret entrances (or destroying a certain borrowed car).

❖ IN THE MUGGLE WORLD:
Dragon's Claw Willow
SALIX MATSUDANA "TORTUOSA"

The Dragon's Claw Willow's gnarled and twisted branches, fondness of shade, aggressive root system, but lack of pugilist prowess make this a safer alternative to the Whomping Willow. It's fast-growing and can get to 30 feet tall and live to around 40 years!

❖ IN THE WIZARDING WORLD:
Mimbulus Mimbletonia

Seen in *Harry Potter and the Order of the Phoenix.* The stinksap-producing, pulsating plant is a rare Assyrian native. It looks like a gray cactus, but with boils.

❖ IN THE MUGGLE WORLD:
Bilberry cactus, whortle-berry cactus, or blue candle, blue myrtle cactus
MYRTILLOCACTUS GEOMETRIZANS

With multiple names, this is also a cactus but without the boils and stinksap. It makes a great houseplant. In its natural setting it grows skyward in a distinctive candelabra shape. If potted, it needs to be repotted every few years to allow for root growth. In the wild, it produces blue berries, a staple in the diets of many indigenous people, including the Apache, Papago, Tewa, and Ute.

EVERY GARDEN IS MAGIC

REAL BUT WEIRD PLANTS

"Personally, these plants have always kind of freaked me out."

—HARRY POTTER,
HARRY POTTER AND THE HALF-BLOOD PRINCE

The Wizarding World isn't the only one with odd and bizarre plants—the Muggle world definitely has its share of greenery that does strange and wonderful things such as folding their leaves when touched or smelling absolutely *awful*. There are plants that look like animals (and we know there are animals that look like plants!). Others have noticeable markings that look like skulls or are shaped like pitchers to catch bugs for dinner. Following are some real but weird plants that would fit right into the Wizarding World as seen in the films. If you choose to use any of these, please note that some are dangerous to animals and humans, so please plant them in a safe, protected place, possibly fenced in, or in pots that pets or any wild animals in your neighborhood can't get to.

Aconitum

Common names: aconite, blue rocket, devil's helmet, leopard's bane, monkshood, mousebane, queen of poisons, wolf's bane, women's bane

Common to the Northern Hemisphere, *Aconitum* is *extremely* poisonous and must be handled very carefully. Its hybrids are horticultural society prize-winning flowers and used by florists. Many varieties of *Aconitum* are the single source of food for many moth caterpillars, like the yellow tiger, purple-shaded gem, and mouse moths.

TOXIC

Actaea pachypoda

Common names: doll's eyes, necklace weed, toadroot, white baneberry, white beads, white cohosh

This poisonous eastern North American native has berries that were once really used as doll's eyes, which bear an eerie similarity to Mad-Eye Moody's eye. All parts of the plant are poisonous, but the berries are the worst. Ingested, the berries lead to cardiac arrest and death. Watch out for them. They're looking for you.

TOXIC

Antirrhinum majus

Common names: bonny rabbits, bulldogs, calf-snout, lion's-mouth, rabbit's mouth, snapdragon, toad's mouth

A Death Eater's delight! The flowers come in a great range of colors, have a citrusy scent, and look like a dragon's head. Once the flowers are spent, things get creepy. The seed heads of snapdragons can look like little human skulls. Lightly pinched, you can even make the flowers' mouths open and close.

Atropa belladonna

Common names: belladonna, deadly nightshade

Belladonna is a toxic perennial plant in the nightshade family Solanaceae, which also includes tomatoes, potatoes, and eggplants. Ingested foliage and berries cause delirium and hallucinations. Witches are believed to use belladonna and other poisonous plants to help them fly to gatherings. Floo powder sounds safer.

TOXIC

Dionaea muscipula

Common name: Venus flytrap

Native to the East Coast of the United States in North Carolina and South Carolina coastal bogs, this carnivorous plant traps its prey in its leaves. Growing in nutrient-poor soil, it feasts on nitrogen-rich crawling insects and arachnids. It tolerates fire well and depends on the occasional fire to eliminate its plant competitors. The Venus flytrap rarely traps its pollinators, as the flowers grow six to ten inches above its leaves.

Dracaena cinnabari

Common name: dragon blood tree

This tree only grows in the granite mountains and limestone plateaus on the island of Socotra, Yemen in the Arabian Sea. Its leaves capture moisture from clouds, sea mist, and rain drizzle and feed it to its roots—the opposite of how most trees work! It has a dark red resin, hence its name. The resin is used as a dye and was used as a source of varnish for 18th-century Italian violin makers.

Hornbeam

Common names: blue-beech, musclewood, ironwood

Found in temperate regions of the Northern Hemisphere, the hornbeam is a smaller tree with very hard wood. The male and female flowers both grow on the same tree and are wind pollinated. The wood is so hard and durable that it gets used for carving boards, tool handles, piano parts, and in years past, as windmill gear pegs.

Dracula simia

Common names: monkey orchid, monkey-like Dracula

It's bananas to think there's an orchid that looks like a capuchin monkey! This southeastern Ecuador native thrives in tropical highland forests and is extremely fragrant. The flowers smell like ripe oranges and blossom any time of the year.

Hydnellum peckii

Common names: bleeding tooth fungus, devil's tooth, redjuice tooth, strawberries and cream

This dessert-looking fungus can be found in North America (mostly the Pacific Northwest), Europe, and more recently in Iran and Korea. The caps "bleed" red droplets that contain a pigment with anticoagulant properties. It is a mycorrhizal fungus, meaning it shares a relationship with the roots of specific host trees, in which it exchanges minerals and amino acids extracted from soil for the fixed carbon from the host. It looks delicious, isn't poisonous, but tastes foul.

Eucalyptus deglupta

Common names: Mindanao gum, rainbow eucalyptus, rainbow gum

Rainbow eucalyptus is a large, fast-growing, evergreen tree native to moist, humid tropical forested areas in New Guinea, Indonesia, and the Philippines. It is the only eucalyptus tree indigenous to the Northern Hemisphere. In its native habitat, it will mature to as much as 250 feet tall. It is best known for its smooth orangish bark which peels to reveal stunning multi-colored bark of pale green, red, orange, gray and purple-brown.

Lithops

Common names: living stones, pebble plants

This member of the ice plant family is native to southern Africa and evolved to blend in with surrounding rocks. It is slow growing and makes a good houseplant given enough sunlight and well-drained soil. Overlooked for generations because it's not easily spotted due to its camouflaging trait, new species are always being discovered.

Ficus altissima

Common names: council tree, lofty fig, strangler fig

The southeastern Asia native can reach up to 100 feet outdoors. Don't worry, it will only reach about 6 feet in your house. It's epiphytic so it usually grows on other trees and puts down aerial roots, like a banyan tree. Like all tropical trees, it likes indirect sunlight and humidity.

Mandragora officinarum

Common name: mandrake
This nightshade plant family member from the Mediterranean region causes delirium and hallucinations. The shape of its roots resembles human figures and has long been used in spiritual practices and magic rituals. It was superstition that anyone who pulls up the root will be condemned to hell and the Mandrake would scream, killing anyone who heard it.

TOXIC

Mimosa pudica

Common names: action plant, sensitive plant, shameplant, sleepy plant, touch-me-not
This Caribbean native folds its leaves when touched, blown, or shaken. This reflex evolved as a defense mechanism to ward off predators or to create shade to reduce evaporation water loss. It also folds at night and opens in daylight. It does flower, but the flowers never last longer than a day.

Origanum dictamnus

Common names: Cretan dittany, dittany of Crete, hop marjoram
Growing only on the mountainside and gorges of the Greek island of Crete, dittany has soft, velvety, white hair and is used for food flavoring, perfumes, and medicines. Because of its growth on mountainsides and cliffs, it was a perilous job collecting the plant—many have died trying. Now it is cultivated in surrounding villages on Crete and is a protected species in its native habitat to protect it from extinction.

Palicourea elata

Common names: girlfriend kiss, hot lips
A tree found in the rainforests of Colombia, Costa Rica, Panama, and Ecuador, *Psychotria elata* has evolved to attract hummingbirds and butterflies. The "lips" are actually only bracts, red leaves, that open to reveal the tree's actual flowers. It is currently endangered because of deforestation. It is widely used as a Valentine's gift where it grows.

Rafflesia

Common names: monster flower, padma raksasa ("giant padma")
This southeast Asian parasitic flower is the largest flower in the world—the largest measured was almost four feet wide. As pretty as it can be, the blossom has a smell of rotten flesh to attract insects, like carrion flies, for pollination. It has no leaves, stems, or roots. It has vines that climb only very specific host trees.

Stachys byzantina

Common names: lamb's ears, wooly hedgenettle
The thick, soft, velvety, silver leaves make this a great and common garden plant. It is easy to grow and touching it is irresistible. It's originally from Turkey, Armenia, and Iran but is now an ornamental plant in cultivated gardens around the world. In Brazil it is used as an herb, prepared battered, deep-fried, sprinkled with lemon juice and said to taste fish-like.

TYPES OF GARDENS

As Professor Sprout teaches her second-year students about Mandrakes in Greenhouse Three, it can be assumed that she grows other plants in other greenhouses for different lessons. Whether a greenhouse, a planter, or a garden, there are many different ways to design a green space that can be beautiful or practical or both. Some gardeners grow related plants, basing their botanical choices on color or smell. Greenery can also serve different functions for their environment, such as gardens assembled to attract certain pollinators or to create a meditative space. There are many more considerations to keep in mind when creating a garden, whether it is the climate you're in, the available space you have, or even if you want to attract butterflies. Every garden is magic and, like the Room of Requirement, can be whatever you need it to be. Following are just a few of the possible choices for your own magical garden.

Balcony or Terrace Gardens

Container gardening where space is limited is also called balcony or terrace gardening. Flowers, vegetables, herbs, shrubs, and trees are all options depending on your space. First, check to see if your building allows plants on balconies! Assess your sunlight (what direction the balcony faces). You'll want to choose plants that thrive in your light conditions: Direct sun all day? Cacti, flowers, vegetables. Shaded? Coleus, hostas, and impatiens.

Butterfly Gardens

Butterfly gardens involve planting flowers that attract butterflies. Find out what butterflies are in your area. Butterfly field guides can help with that. Butterfly gardens are best placed where butterflies will have sun, water, and shelter. Each species of butterfly has a different host plant. Some quick online research will tell you what host plants to plant for the butterfly species you want to attract and encourage.

Cutting Gardens

A cutting garden is where you grow flowers for bouquets and arrangements. Cut flowers are what you see at a florist's shop. They are used for decorative purposes in garlands, bouquets, flower arrangements, floral baskets, and wreaths. When you have them in your own garden, you can use them in vases.

Fairy Gardens

A fairy garden is a small container garden or a small space near a tree with miniature-size plants and elements, the purpose of which is to be a place for fairies to live. The habitat can be built so that fairies will visit when no one is looking.

Herb Garden

An herb garden is an outdoor garden used solely to grow herbs for seasonings, flavorings, medicinal uses, and aromatherapy. They can be any size or shape and can contain different herbs. An herb garden may take up an entire yard or may simply be planted in a small window box container.

Labyrinth Gardens

Labyrinth gardens are meant as a place for personal contemplation and renewal. In a labyrinth, one follows a curving pathway leading to a center. It is not a maze, which has fake paths and dead ends. Labyrinths are not designed to be difficult. Once at the center, you take the same path out. In the Middle Ages, labyrinths were made on the floors of religious buildings. One of the most ancient labyrinths is built right into the floor of the Chartres Cathedral in France; it was meant to provide a meditative journey for body and spirit.

Native Plant Gardens

Native plant gardens are the most environmentally friendly gardens and good for all native creatures. The native plants, insects, and animals where you live have evolved together over eons and are part of a local ecosystem. They sustain each other. By planting native plants, you become part of that ecosystem, too. With non-native plants and loss of habitat by humans, your yard and garden (or rooftop or balcony) can create a haven for birds and pollinating insects.

Organic Gardens

Organic gardens usually contain vegetables and herbs, but more often now also feature ornamental gardens. The main factor that makes a garden organic is that it does not use artificial or processed fertilizers and also does not use inorganic pesticides for insects or weeds. Organic gardening improves soils.

Pollinator Gardens

Pollinator gardens support and maintain pollinators by supplying food in the form of pollen and nectar to ensure that the pollinators stay in an area to pollinate crops for fruit and vegetable production.

Sensory Gardens

A sensory garden is designed to stimulate the senses. Stimulation occurs courtesy of plants and materials that engage one's senses of sight, smell, touch, taste, and sound. These gardens are beneficial to children and adults, especially those who have sensory processing issues, like autism and other disabilities.

Vegetable Gardens

Vegetable gardening includes selecting a site, planning the garden, preparing the soil, choosing the seeds and plants, planting a crop, and nurturing the plants until they are ready for harvest. The end result is fresh produce to eat, share, or sell.

Vertical Gardens

A vertical garden is a vertically suspended panel on which plants are grown typically using hydroponics. These unique structures can either be freestanding or attached to a wall and go by different names: living green walls, plant walls, and moss walls just to name a few. Vertical structures of plant life can be as small as a picture frame or massive enough to cover an entire wall.

THE MAGIC OF POLLINATORS

A pollinator is any creature that helps distribute pollen from one plant to another. Pollen needs to go from a male flower to a female flower. That movement, from stamen to stigma, enables a plant to fertilize and produce fruits, seeds, and young plants.

Pollinators include bees, butterflies, birds, beetles, and bats, as well as wasps, moths, insects, and small mammals. While they're enjoying a flower's nectar, in search of food, shelter, or nest-building materials, they inadvertently transfer pollen that sticks on their bodies from one plant to another. Bees are super pollinators; food crop production is dependent upon them to pollinate.

Providing a safe habitat for pollinators to live and reproduce is important to any ecosystem. More than 70 percent of all flowering plants on earth are pollinated by creatures. One out of every three bites of food you eat exists because of pollinators!

Want to provide a safe habitat for pollinators? Like we do, they need food, water, shelter, and space. Common to all pollinators are flowering plants and different bloom times from spring to fall, providing food and nesting space. Also common to all pollinators is water. A shallow tray with water and pebbles (like the Pensieve Birdbath on page 145) provides a safe landing spot for birds, bees, and butterflies. The most important thing to remember is to NOT use pesticides, weed killers, or fertilizers with synthetic chemicals added. They are Death Eaters for pollinators.

Different pollinators do require different habitats. Most require specific plants. Some like open areas, but others don't. You can design a pollinator garden for bees, or hummingbirds, or butterflies. Some quick research will yield the types of plants and flowers each needs for food and shelter.

WHY USE LATIN FOR SPELLS, CHARMS, AND PLANT NAMES?

For spells, charms, enchantments, hexes, and curses, it's important to get to the truest meaning of the incantation. Think of Harry Potter mispronouncing "Diagon Alley" the first time he uses Floo powder, and finding himself in the dangerous Knockturn Alley due to his mistake. A "universal" language (and correct pronunciation) can make all the difference, and so for the spells, Latin, the mother of all European languages, forms the basis for naming magical charms and curses. This important universality serves the same purpose for plant names.

Latin is a "dead" language, in that no new words get added to it over time. Its words convey precise meanings that do not change, whether you normally speak English or French or Spanish. So, using Latin for botanical naming conventions and writing ensures that there's no doubt what plant you're talking about.

Carl Linnaeus (1707–1778) came up with "binomial nomenclature" in 1753, where each species of plant has a name that has two parts. Each part of the name used is Latin.

- The first word is the name of the genus. The genus is the kind, or group marked by common characteristics.

- The second word is the name of the species, which describes in more detail the individual characteristics of the plant belonging to the genus.

- Sometimes there's a third part below the rank of species—a variety or cultivar (a cultivar is a manmade variety).

Common names, like black-eyed Susans, coral bells, and cone flowers are all either too general a term or are called one name in one part of the country and a different name in another. Using one language for plant nomenclature assures botanists, scientists, growers, retailers—and you—are all talking about the same plant.

ZONE GROWN

The U.S. Department of Agriculture Plant Hardiness Zone Map (a Marauder's Map for plants) is a standard gardeners can use to determine what plants will most likely thrive outdoors where they live. It's based on average minimum winter temperature. The U.S. is divided into 13 zones, and broken further down, in 10°F increments from Hawaii's hot Hilo at 13a to Alaska's cool 1a North Slope. Each state will have many hardiness zones within it.

Many factors contribute to what plants can grow where you live. Microclimates, where weather conditions differ from surrounding areas, can be found all over—from tiny areas of a few feet to as much as a few square miles. Microclimates might include near temperature-moderating bodies of water, urban settings where buildings and infrastructure absorb and retain the sun's heat and divert winds, and mountainous areas where average temperatures lower with altitudes.

Growing plants from adjacent higher zone numbers in your own garden successfully is called Zone Pushing. Microclimates and special overwintering care, like piling mulch over a plant, can make you a Zone Pusher.

All countries have their own versions of plant hardiness zones, most based on average winter temperatures, many following the U.S. standards for hardiness zones. To find out what U.S. Zone you live in, all you need is your zip code and to visit PlantHardiness.ars.usda.gov.

NATIVE PLANTS

A plant is native to a region because it has evolved along with native wildlife and soil and has developed a symbiotic relationship over thousands of years. If a plant occurs naturally in a region, without human introduction, it is considered native. Native plants to North America are recognized as occurring prior to European settlement.

A native plant to you may not be a native plant in another part of the country. There are plenty of non-native plants growing in gardens and in the wild near you—and they may thrive—but they are not serving the native wildlife as well as native plants. Some non-native plants may even be invasive in your area (but not in their native habitat) and cause harm to your local ecosystem as they take over space and nutrients native plants could use.

A native plant garden, or native plants in a garden along with non-natives will attract pollinators and beneficial insects and help *them* thrive.

Published by Titan Books, London, in 2023.

TITAN BOOKS

A division of Titan Publishing Group Ltd
144 Southwark Street
London SE1 0UP
www.titanbooks.com

Find us on Facebook: www.facebook.com/TitanBooks
Follow us on Twitter: @titanbooks

Copyright © 2023 Warner Bros. Entertainment Inc. WIZARDING WORLD characters, names and related indicia are © & ™ Warner Bros. Entertainment Inc. WB SHIELD: TM & © WBEI. Publishing Rights © JKR. (s23)

All rights reserved.

Published by arrangement with Insight Editions, San Rafael, California.

www.insighteditions.com

No part of this publication may be reproduced, stored in a retrieval system, or transmitted, in any form or by any means without the prior written permission of the publisher, nor be otherwise circulated in any form of binding or cover other than that in which it is published and without a similar condition being imposed on the subsequent purchaser.

A CIP catalogue record for this title is available from the British Library.

ISBN: 9781803367057

Publisher: Raoul Goff
VP of Licensing and Partnerships: Vanessa Lopez
VP, Creative: Chrissy Kwasnik
VP, Manufacturing: Alix Nicholaeff
VP, Editorial Director: Vicki Jaeger
Publishing Director: Jamie Thompson
Designer: Lola Villanueva
Editor: Anna Wostenberg
Editorial Assistant: Sami Alvarado
Managing Editor: Maria Spano
Senior Production Editor: Katie Rokakis
Production Associate: Deena Hashem
Senior Production Manager, Subsidiary Rights: Lina s Palma-Temena

Photographer: Ted Thomas
Assistant Photographer: Amani Wade
Stylist: Elena Craig
Assistant Stylist: Patricia Parrish
Plant Aficionado: Meredith Law
Photoshoot Art Director: Judy Wiatrek Trum

Technical Illustrator: Anastasia Shumeeva
Cover illustrator: Paula Hanback

Manufactured in China

10 9 8 7 6 5 4 3 2 1